Green Alternatives and National Energy Strategy

Green Alternatives and National Energy Strategy

The Facts behind the Headlines

PHILIP G. GALLMAN

The Johns Hopkins University Press

Baltimore

The Johns Hopkins University Press
2715 North Charles Street
Baltimore, Maryland 21218-4363
www.press.jhu.edu

Library of Congress Cataloging-in-Publication Data

Gallman, Philip G.
Green alternatives and national energy strategy : the facts behind the headlines /
Philip G. Gallman.
    p. cm.
Includes bibliographical references and index.
ISBN-13: 978-1-4214-0197-3 (hardcover : alk. paper)
ISBN-10: 1-4214-0197-5 (hardcover : alk. paper)
1. Power resources—United States. 2. Fossil fuels—Environmental aspects—United
States. 3. Energy policy—United States. 4. Energy resources development—United
States. I. Title.
TJ163.25.U6G35 2011
333.790973—dc22   2010052919

A catalog record for this book is available from the British Library.

*Special discounts are available for bulk purchases of this book. For more information,
please contact Special Sales at 410-516-6936 or specialsales@press.jhu.edu.*

The Johns Hopkins University Press uses environmentally friendly book materials,
including recycled text paper that is composed of at least 30 percent post-consumer
waste, whenever possible.

All we want are the facts, ma'am.

All we know are the facts, ma'am.

—Sergeant Joe Friday, *Dragnet*

Contents

Figures and Tables

....

Figures

Tables

Preface

The world, especially the United States and other industrialized countries, is facing a growing energy crisis. On the one hand, demand for energy is increasing because of the dual pressures of growing world population and growing demand for industrialization on the part of that population. On the other hand, limited reserves of conventional sources of energy are running out. On top of that, increasing industrialization and consumption of conventional energy sources are producing increasing levels of atmospheric pollutants and greenhouse gases, causing worldwide climate change and global warming. If these trends continue, they could have catastrophic effects on human life on this planet.

In addition, the United States depends on foreign, not necessarily friendly, nations for a significant proportion of its energy. Production and export decisions by the nations that make up the Organization of Petroleum Exporting Countries (OPEC) can, and several times have, thrown the US gasoline supply into disarray with increased prices, diminished supply, and long lines at filling stations. Being at the mercy of foreign nations is intolerable.

Most people acknowledge that we need to develop a long-range energy plan that addresses three key goals:

- Become independent of foreign oil.
- Reduce greenhouse gases and pollution.
- Develop renewable clean sources of energy.

Many people believe that the solution to all three problems is to eliminate or at least reduce demand for gasoline by developing alternative (i.e., "green") vehicles and to develop renewable, nonpetroleum sources of energy, such as solar power and wind power. While nothing is wrong with this viewpoint, it is overly simplified. My purpose in writing this book is to guide the reader through what we know about alternative vehicles and energy and explain the more important benefits, problems, issues, decisions, and trade-offs that lie before us.

Engineers are trained to analyze and evaluate hard data, to follow where the data lead, and to present their results without bias or opinion. When data are incomplete or missing, that fact should be mentioned. When opinion is requested, good engineering practice is to explain the reasoning behind the opinion and to clearly indicate that it is opinion. I have followed my engineering training in developing this book. My goal is to guide the reader through the basic data and explain what they mean and to present the data and the logical analysis clearly enough that readers, whether they agree or disagree with my conclusions, understand the facts rather than the hype.

Raw, unadulterated data are crucial to this undertaking, especially data concerning production, consumption, importation, and reserves of natural resources. The best sources have been government databases, and the Internet has proved to be an invaluable research resource because a great deal of raw data are available to the reader easily and in usable format. Unfortunately, websites tend to be ephemeral, as pages may be updated and changed unexpectedly, and so they are not the best references. The approach I have followed to provide dependable references to government databases is to give Internet hyperlinks on this book's companion website (greenalternativesandenergy.com) and to cache pages so that the original is available even if the website is changed.

I started my investigation by looking at electric cars because I am an electrical engineer. It soon became evident that one cannot discuss electric cars productively without comparing and contrasting them with the other alternatives: hybrids, natural gas, and so on. Once I included the alternatives in my study, it became clear that one must also examine the effect of "green" vehicle technology on our energy supply and consumption. New energy technology and new sources of energy such as wind and solar power and how they relate to vehicle energy consumption needed to be folded into the analysis. My investigation led to several unexpected and surprising observations reinforcing my belief that most people are misinformed about energy and green automobiles. The situation is much more complex than most people realize. The better informed we are, the better we can withstand the urgings of shortsighted special-interest groups.

All of our gasoline and diesel fuel comes from oil, so reducing or eliminating demand for gasoline and diesel would certainly have a major influence on our need to import foreign oil. However, a large percentage of the oil we consume goes to products other than vehicle fuel, products we cannot live without, such as fuel oils, lubricating oil, and feedstock for the petrochemical industry. Becoming independent of foreign oil is much more complicated than improving the fuel efficiency of automobiles. Independence from foreign oil requires finding substitutes or alternative sources of feedstock for numerous indispensable products other than gasoline.

Vehicles are a major source of pollution and greenhouse gases, so reducing demand for gasoline and diesel would certainly reduce emissions of greenhouse gases and pollution. The electric car is a promising solution. However, conventional power plants that burn hydrocarbon fuels produce more pollution and greenhouse gases than burning gasoline does, and most of our electricity comes from burning fossil fuels, particularly coal. We cannot effectively evaluate electric cars without addressing the generation of electricity. Advanced battery technology, though important, is not the primary issue. Developing clean sources of electricity is much more critical.

The primary renewable sources of energy—solar power and wind—certainly provide clean and inexhaustible electricity. However, both demand a huge real-estate footprint, are decidedly undependable, require massive expansion of the nation's network of unpopular transmission lines, and jeopardize the stability of the power grid. The cost and efficiency of photovoltaic cells are the least of the problems.

There is heated discussion surrounding how much fossil fuel we have left in the United States and worldwide. The question is not really how much is in the ground per se but how much we can extract at acceptable cost. It is as much a question of economics as it is a matter of geology. Different analysts present different views of reserves, how much of a particular natural resource exists, and how much it would cost to extract it. They all agree, however, that the conventional natural resources are finite and that we are rapidly depleting them. They differ only in their timelines.

Rising demand and decreasing supply is a worldwide problem. I do not dwell on the worldwide problem in this book because I am more concerned about the national problem we face in the United States. We are not a developing third-world country, we have a unique situation with regard to supplies of natural resources, we have a uniquely American relationship with automobiles and travel, and so on. Similarly, I do not discuss how individual homeowners can conserve energy because I am more interested in the large-scale, national issues. For example, it is quite possible for homeowners to power their homes with solar power. All you need is a location with good sun and a good supply of storage batteries to take care of demand when the sun is not shining. However, what the individual can do does not necessarily work on a national level. Running the whole country on batteries at night is not practical. Similarly, much of the "living green" literature and discussion, though very useful to individuals, does not apply at the national level.

We have serious challenges ahead. Oil will run out, even if we do not know exactly when. If we do not prepare for that event, the result could be very painful. We will need a Herculean effort along with massive amounts of research and development funding. With limited resources, both labor and money, and rapidly depleting oil supplies, we had better get the priorities right. From what I see in the headlines, fifteen-second sound bites, and cable news reports, the majority of US citizens are woefully underinformed or misinformed about what we need and what technology can or cannot do for us. I hope this book helps improve the situation.

In chapter 1, "Conventional Energy Sources," I review the primary sources of energy—petroleum, natural gas, coal, and nuclear power—from the point of view of how we get the energy, transport it, store it, and use it. I cover electricity even though it is not a primary source of energy or a natural resource because it is generated using natural resources. I show how much of each type of energy we use in the United States and the benefits and liabilities of how each energy source is stored and transported. I encountered some surprises:

- Simply stopping gasoline consumption will not enable us to stop importing oil.

- We do not have unlimited natural gas resources. Indeed, we are importing significant amounts already and are on track to run out of natural gas not much later than when we run out of oil.

- Generating electricity is by far much more polluting than using gasoline in automobiles. We would do much more for the environment by cleaning up the supply of electricity than by cleaning up automobile engines.

Chapters 2 and 3 address automobiles. Chapter 2, "Conventional Vehicles," opens with a discussion of the internal combustion gasoline engine and standard drivetrain in order to identify where energy losses occur and where we can improve fuel efficiency. Discussions of the current candidates for alternative vehicles follow in chapter 3, "Green Vehicles": diesel, flex-fuel, natural gas, hybrid electric, series electric, plug-in electric, and hydrogen fuel cell. I discuss the engineering behind each alternative and the benefits and drawbacks of each technology. The chapter closes with my analysis of what should work and what alternatives we should pursue and which we should abandon. Here again there were surprises:

- Most of the up-and-coming green vehicles increase overall cost so much that the financially sound choice for an individual is to ignore them.
- Widespread conversion to "fuel-efficient" diesel vehicles would actually increase demand for oil.
- Natural gas is a questionable source of fuel for private automobiles.
- Switching to plug-in electric vehicles would significantly increase pollution and overload the national power grid unless generating capacity is markedly expanded with clean sources.

Chapter 4, "Green Energy Sources," discusses renewable resources. Growing population; increased usage of electric and electronic tools like computers; recognition that coal, a major source of electricity, is extremely polluting and damaging to the environment; and the need to deploy electric cars point strongly in the direction of clean, renewable sources of electricity such as solar and wind power. Here the issue is not so much the physics or even the cost of generating electricity but rather the ability to distribute electricity efficiently to the entire nation from the very limited number of sites with adequate solar or wind power potential. The second issue is providing electricity when the sun does not shine or the wind does not blow. Temporary energy storage capable of running the entire country for hours or days seems implausible. Geothermal and hydroelectric power sources are less erratic than solar and wind power but require the right geological conditions. The number of practical locations for new geothermal and hydroelectric power plants is limited. Nuclear power, a clean and potentially inexhaustible source of energy, is a possible solution. I discuss the promise of "clean coal" technology for generating electricity and the use of Fischer-Tropsch processes to produce gasoline and other products from coal and natural gas rather than from oil. Some of my conclusions were sobering:

- Neither solar nor wind power will provide a significant portion of our electricity demand.
- Moderate development of solar and wind power may jeopardize the national power grid.
- Clean coal technology is potentially more dangerous than storing waste from nuclear power plants.

My conclusions are presented in chapter 5. When one completes a thorough transportation and energy-system analysis as I have done, it is clear that the United States has a serious energy problem and that the future is fraught with challenges. However, the correct path forward is inescapable and obvious. We need to replace the current energy and petrochemical paradigm based on fossil fuels with something else based on clean and inexhaustible resources. Moreover, we need to do so soon, before we run out of oil. That does not give us much time. The worldwide situation makes our problem more difficult. I see a collision between increasing population and industrialization in the third world, on one hand, and finite natural resources, on the other. I am concerned that the situation is worsening and we are not responding quickly enough or correctly.

Several themes recur throughout this book. Please keep them in mind as you read. First, both oil and natural gas will run out. I do not know exactly when, but it is highly probable that it will happen during the lifetimes of the children born today or their children. Severe disruption will occur if we do not prepare adequately. Moreover, energy-saving methods that work well for individuals do not necessarily work at the community level, and methods that work well for small communities do not necessarily work at the national level. Many of the green technology success stories in the popular press are strictly at the individual level and have little influence nationally. Finally, oil provides many indispensable products other than gasoline and diesel. Reducing demand for gasoline will have little effect on demand for oil unless we find alternative sources for the other products.

My intent in writing this book is to point out the major facts and issues, many of them seemingly ignored or misunderstood, in the hope that, being better informed, we can make good decisions about our energy future. I present what I believe are the critical engineering facts about energy and green vehicles. I put forward my own ideas about what we should do and what we should not continue pursuing. I may be wrong. I may have missed something. We can only try to do our best given our current knowledge and the developing world energy situation. If this book helps you understand the issues and make good decisions, I will have succeeded in what I set out to do.

I want to thank Mollie Wisecarver for her valuable comments and careful editing of the early manuscript and George Roupe for his meticulous copy-editing. I also thank my wife, Minnie, for her ongoing support and continuous help keeping the manuscript on track. This book would not have come to be without these people.

Abbreviations

AC	alternating current
ARS	acute radiation syndrome
Btu	British thermal unit
CAFE	Corporate Average Fuel Economy
CCS	carbon capture and sequestration
CNG	compressed natural gas
CO_{2e}	carbon dioxide equivalent
CSP	concentrating solar power
dB	decibel
DC	direct current
DOE	Department of Energy
E10	fuel consisting of 10% ethanol and 90% gasoline
E85	fuel consisting of 85% ethanol and 15% gasoline
EERE	Office of Energy Efficiency and Renewable Energy
EIA	Energy Information Administration
EPA	Environmental Protection Agency
EV	electric vehicle
FLL	fertilizer, livestock flatulence, and landfills
gge	gallon of gasoline equivalent
GVWR	gross vehicle weight rating
GWP	global warming potential
HEV	hybrid electric vehicle
HFC	hydrofluorocarbon
HFCV	hydrogen fuel-cell vehicle
H-PEV	hybrid plug-in vehicle
Hz	hertz
ICEV	internal combustion engine vehicle

IGCC	integrated gasification combined cycle
kg	kilogram
kWh	kilowatt-hours
L	liters
LNG	liquefied natural gas
LPG	liquefied petroleum gas
MJ	megajoules
MMS	Minerals Management Service
MMt	million metric tons
mpg	miles per gallon
mpgge	mpg gasoline equivalent
mph	miles per hour
m/s	meters per second
MTBE	methyl tertiary butyl ether
MW	megawatts
MWh	megawatt-hours
NaS	sodium-sulfur
NGV	natural gas vehicle
NHTSA	National Highway Traffic Safety Administration
NREL	National Renewable Energy Laboratory
OPEC	Organization of Petroleum Exporting Countries
PEV	plug-in electric vehicle
P-PEV	pure plug-in electric vehicle
psi	pounds per square inch
PV	photovoltaic cell
quad	quadrillion Btu
R&D	research and development
rpm	revolutions per minute
SEV	series electric vehicle
SPL	sound pressure level
SUV	sport-utility vehicle
Tcf	trillion cubic feet
USGS	United States Geological Survey
UTRR	undiscovered technically recoverable reserves

Green Alternatives and National Energy Strategy

CHAPTER 1

Conventional Energy Sources

The immediate goals of our national energy strategy should be to reduce our dependence on foreign sources of energy, to move away from dependence on limited natural resources and toward renewable sources, and to reduce emissions of pollutants and greenhouse gases. The chapters of this book discuss what we can do with vehicles to achieve these goals and what novel or alternative sources of energy are available. To understand the practicality and effects of these alternatives, one must first understand the current situation with regard to our major conventional energy sources. By major conventional energy sources, I mean oil, natural gas, coal, and nuclear power. I discuss alternative energy sources, such as wind, solar, geothermal, and hydroelectric, in chapter 4.

US Energy Consumption

The US Energy Information Administration (EIA) provides details of US consumption of primary and secondary energy sources.[1] By *primary energy source*, I mean energy that we get directly from nature: oil, natural gas, coal, and uranium. In contrast, a *secondary energy source* is one that we manufacture from a primary source. In particular, we generate electricity by burning some primary fuel, most commonly coal and natural gas. The unit of energy *quad*, short for quadrillion British thermal units (Btu),[2] is widely used in the energy industry but is probably unfamiliar to readers outside the industry. I express some numbers in quads to simplify comparisons with the energy literature, but the kilowatt-hour (kWh), a standard unit of electrical energy, is more useful for this book because of its focus on electric cars and my belief that electricity will eventually become the dominant source of energy for the consumer.

Table 1.1 and figure 1.1 show our annual consumption of energy and the percentage of our annual consumption provided by each primary source. More important, the table also shows the amount of each primary source consumed. We measure petroleum in barrels (42 gallons per barrel); natural gas in cubic feet, usually in trillions of cubic feet (Tcf); coal in short tons;[3] and nuclear reactor fuel in pounds of uranium oxide. The raw amount figures are useful as reference points when we discuss the effect of different automobile fuels on national resource usage and pollution emission.

Renewable energy sources, which contribute very little to the overall energy supply, are further broken down as shown in table

TABLE 1.1 *US Consumption of Primary Energy Sources, 2008*

Source	Energy		Percentage of total energy consumption	Raw amount
	quads	trillion kWh		
Petroleum	37.1	10.9	37.4	7.1 billion barrels
Natural gas	23.8	7.0	24.0	23.2 Tcf
Coal	22.4	6.6	22.6	1.1 billion short tons
Uranium (U_3O_8)	8.5	2.5	8.51	51.3 million pounds
Renewable	7.3	2.1	7.4	N/A
Total	99.1	29.1	100.0	

Sources: Energy by source: EIA, "Annual Energy Review, 2009," table 1.3, "Primary Energy Consumption by Source, 1949–2009," www.eia.doe.gov/emeu/aer/txt/ptb0103.html. Raw amounts: EIA, "Annual Energy Review, 2009"; petroleum: table 5.1, "Petroleum Overview, Selected Years, 1949–2009," www.eia.doe.gov/aer/pdf/pages/sec5_5.pdf; natural gas: table 6.1, "Natural Gas Overview, Selected Years, 1949–2009," www.eia.doe.gov/aer/pdf/pages/sec6_5.pdf; coal: table 7.1, "Coal Overview, Selected Years, 1949–2009," www.eia.doe.gov/aer/pdf/pages/sec7_5.pdf; uranium: table 9.3, "Natural Uranium Overview, Selected Years, 1949–2009," www.eia.doe.gov/aer/pdf/pages/sec9_7.pdf.

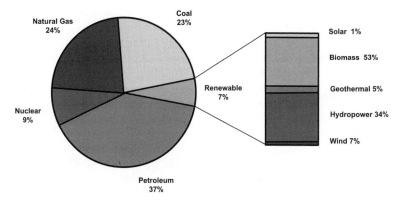

FIGURE 1.1 US Consumption of Energy by Source, 2008. Total US energy consumption was 99.1 quads (29 trillion kWh) in 2008. Consumption of renewable sources was 7.3 quads (2.1 trillion kWh). The bar chart at right shows percentages of renewable energy obtained from each source.

Source: EIA, "Annual Energy Review, 2009," table 1.3, "Primary Energy Consumption by Source, 1949–2009," www.eia.doe.gov/emeu/aer/txt/ptb0103.html.

1.2 and the bar graph on the right-hand side of figure 1.1. Hydroelectric dams and biomass make up most of the renewable sources at the current time. The current darlings of renewable energy sources, solar and wind, are now insignificant sources of electricity. I discuss in chapter 4 whether they and geothermal and hydroelectric power can become major contributors.

These tables show the ultimate sources of our energy. They do not indicate where our energy goes. This is shown in figure 1.2.

This figure shows where energy is consumed: transportation, industry, residential and commercial, and electric power. It summarizes a great deal of data and it takes a little effort to understand. The ellipses on the left list the primary energy sources and how much each contributes to overall US consumption. For example, coal provided 22.5 quads of the 99.1-quad total energy consumption in 2008. The rectangles on the right show usage by sector. For example, electric power generators consume 40.1 quads of the 99.1-quad total. The numbers at each end of the lines

TABLE 1.2   *US Consumption of Renewable Energy Sources, 2008*

Source	Energy quads	Energy billion kWh	Percentage of renewable energy consumption
Solar	0.1	27	1.3
Biomass	3.9	1,138	53.2
Geothermal	0.4	105	4.9
Hydroelectric	2.4	718	33.6
Wind	0.5	151	7.1
Total	7.3	2,139	100.0

Source: EIA, "Annual Energy Review, 2009," table 1.3, "Primary Energy Consumption by Source, 1949–2009," www.eia.doe.gov/emeu/aer/txt/ptb0103.html.

connecting the ellipses and rectangles indicate the percentage of each source that goes to each sector and the percentage of each sector of consumption supplied by each source. For example, the number 91 near the coal ellipse on the line from coal to electric power indicates that 91% of our coal goes to generating electricity. The rest of coal goes to industry and home heating. The number 51 on the other end of the same line indicates that 51% of electricity is generated using coal. The line connecting the Nuclear ellipse to the Electric Power Generation rectangle indicates that all nuclear power goes to generating electricity, while 21% of our electricity comes from nuclear power.

These tables and figures tell almost everything one needs to know about our consumption of energy-related natural resources, but they do not give the details of our consumption of secondary energy sources—gasoline, diesel, and electricity—which are important sources of energy for the end consumer but which are manufactured from primary natural resources, such as oil and coal. Secondary energy consumption is shown in table 1.3. These data points will become important in chapter 3, where I discuss alternative vehicles.

The reader should beware numbers dealing with consumption of electricity. Table 1.3 shows 4.1 trillion kWh (14 quads) of electricity *consumed by the end user.* Figure 1.2 shows 11.8 trillion kWh

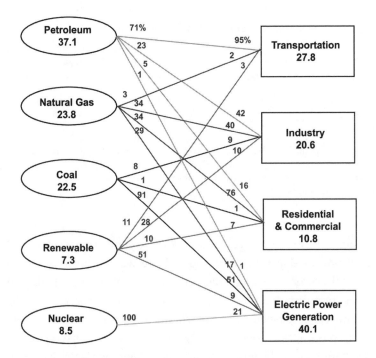

FIGURE 1.2 US Energy Sources versus Consumption Sectors, 2008. The ellipses on the left list the primary energy sources and how much each contributes in quads to overall US consumption. The rectangles on the right show usage by sector. The numbers beside the lines connecting ellipses and rectangles indicate the percentage of each source that goes to each sector of consumption (the numbers near the ellipses) and the percentage of each sector supplied by each source (the numbers near the circles). The differing total values on the left and right reflect independent rounding.

Source: Redrawn from illustration at EIA, "Annual Energy Review, 2009," www.eia.doe.gov/emeu/aer/contents.html.

(40.1 quads) of *primary energy source* consumed to produce electricity. The numbers are different because the usual method of generating electricity, burning a fuel to generate heat that drives a steam turbine that drives a generator, is only about one-third efficient on average. That is, it takes 11.8 trillion kWh of raw resources to produce 4.1 trillion kWh of electric power delivered to the consumer.

TABLE 1.3   *US Consumption of Secondary Energy Sources, 2008*

| Source | Energy | | Raw amount |
	quads	trillion kWh	(billion gallons)
Gasoline	17.3	5.1	138
Diesel	8.5	2.5	61
Electricity	14.0	4.1	N/A

Sources: Gasoline and diesel: EIA, "Petroleum Navigator," http://tonto.eia.doe.gov/dnav/pet/pet_cons_psup_dc_nus_mbbl_a.htm; electricity: EIA, "Annual Energy Review, 2009," table 8.1, "Electricity Overview," www.eia.doe.gov/emeu/aer/pdf/pages/sec8_5.pdf.

(Renewable sources of electricity—hydro, geothermal, solar, and wind power—have different issues. Here we are concerned with thermal power plants and the consumption of primary fuels and production of pollution and greenhouse gases.) When reading the literature, one must be clear as to whether the consumption numbers mean consumption of electricity by the consumer or consumption of raw resources by the power station generating the electricity. I will be careful to make that distinction whenever I refer to electricity. Generating electricity with turbines is inherently inefficient.

US Greenhouse Gas Emissions

Air pollution has been a serious health concern ever since the industrial revolution. The set of specific pollutants we focus on changes with the decades. Currently, the primary concern is with greenhouse gases because of their effect on global warming. Because of this concern, I emphasize greenhouse gases, which make up a subset of air pollution.[4] Nonetheless, I discuss traditional pollution later in this chapter.

I am interested in greenhouse gases because of their role in producing global warming. There has been a great deal of discussion of global warming: whether or not the planet's average temperature is rising and whether or not the temperature increase is due to man's industrial activity. As of this writing, there is a consensus that global warming is real and man's production of greenhouse gases is partially responsible. This might change in the future, but reducing greenhouse gases to protect the planet and ourselves is a worthy goal in any case.

The greenhouse gases considered to cause global warming are carbon dioxide (CO_2), methane (CH_4), nitrous oxide (N_2O), and gases with high global warming potential (GWP), such as hydrofluorocarbons (HFCs), perfluorocarbons, and sulfur hexafluoride (SF_6). The amount of gas produced is expressed by weight in million metric tons (MMt). Carbon dioxide is by far the most prevalent greenhouse gas, and it is customary to tabulate emissions of the other greenhouse gases in terms of carbon dioxide equivalent, that is, the amount of carbon dioxide that would have the same effect as the actual amount of the particular gas. Total US greenhouse gas emission in 2007 was 7,282.4 MMt. Over 80% was CO_2. The rest was methane, nitrous oxide, and high-GWP gases.[5]

The Intergovernmental Panel on Climate Change establishes the GWP factor for each gas.[6] For example, the GWP factor for nitrous oxide is 298, which means that a metric ton of nitrous oxide has the same effect on global warming as 298 metric tons of carbon dioxide. One of the ironies of trying to manipulate nature is that HFCs, which were introduced in the 1990s to replace the ozone-layer-depleting gases then used in air conditioners, refrigeration, and insulating foam, have turned out to be high-GWP gases and more damaging to the environment than the gases they replaced.[7]

Figure 1.3 shows the US emissions from individual sources of energy in 2007. Total emissions amounted to 7,282 MMt of CO_2. Oil accounted for 2,580 MMt, and generating electricity was responsible for 2,433 MMt. Fertilizer, livestock flatulence, and landfills (FLL) is methane produced by agriculture, farming livestock, and human garbage landfills (not human sewage). The "other" category includes greenhouse gases other than CO_2 and methane and emissions from various industrial processes.

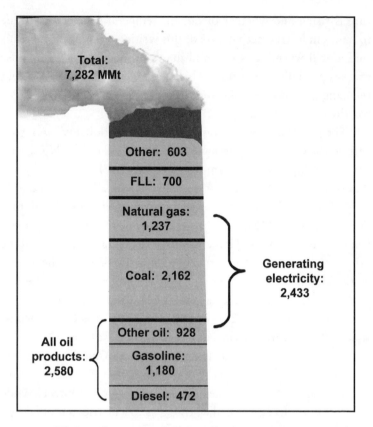

FIGURE 1.3 US Greenhouse Gas Emissions by Source, 2007 (MMt of carbon dioxide equivalent). Total annual emissions were 7,282 MMt. Emissions from oil totaled 2,580 MMt, split among diesel, gasoline, and other products. Generating electricity consumed virtually all coal, 30% of natural gas, and 2% of oil and produced a third of all greenhouse gas emissions (2,433 MMt). Motor fuel—gasoline and diesel combined—produced less than a quarter of the total. Generating electricity in power plants is a much greater source of greenhouse gases than automobiles (1,652 MMt). Fertilizer, livestock flatulence, and landfills (FLL) produced almost 10% of the greenhouse gases, nearly twice the amount produced by diesel fuel. One might argue that we could do more for the environment by reducing the use of fertilizer in farming and by reducing consumption of meat and dairy than by improving diesel engines.

Source: EIA, "Emissions of Greenhouse Gases Report," December 2008, www.eia.doe.gov/oiaf/1605/archive/gg08rpt/index.html.

Figure 1.3 illustrates three very interesting points. First, generating electricity produces twice the CO_2 that gasoline does for the same amount of useful energy. Perhaps we should be more concerned with electric power plants than with automobiles, at least where CO_2 is concerned. Second, natural gas is responsible for more total greenhouse gas than gasoline, though slightly less than gasoline and diesel combined. This is because of the large amount of natural gas consumed rather than because of any inherent "dirtiness." Finally, when we look at FLL, we see that farming and processing human garbage, somewhat surprisingly, generate more greenhouse gas than does burning diesel fuel. As populations grow, the amount of such gas will increase. Based on these data, one might argue that shifting to organic farming and vegetarianism would have greater positive effect on global warming than taking all the diesel cars off the road. I am not suggesting that we all become vegetarians, but I want to introduce the idea that we sometimes are short sighted and concentrate on issues other than the main problem. In the same vein, underground coal-seam fires and peat fires come to mind as unexpected nonindustrial sources of greenhouse gases. Underground coal-seam fires in China produce almost as much CO_2 as all the diesel cars in the United States. In Indonesia, deforestation has led to drying out of the peat cover, which in turn has led to peat fires. Peat fires in Indonesia in 2006 produced more CO_2 than US gasoline and diesel use combined.[8]

Figure 1.4 shows the contribution of the main fuels to greenhouse gases in relationship to the energy we get from them, that is, pounds of greenhouse gas per 1,000 kWh of energy extracted. At one end of the spectrum of major fossil fuels, natural gas is clearly the cleanest source of energy. Renewable resources, such as wind and solar power, are the cleanest of all, but among fossil fuels, natural gas is the winner. At the other end of the fossil-fuel spectrum, coal produces almost twice the greenhouse gas that natural gas does for the same energy content. Compounding the problem is that almost all coal is used to generate electricity and that coal-fired turbine generators convert only about a third of the energy

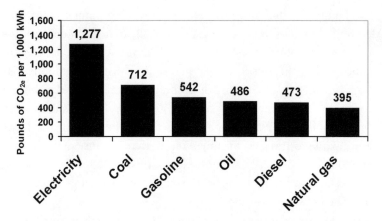

FIGURE 1.4 US Greenhouse Gas Emissions by Energy Content, 2007. Natural gas clearly produces the least greenhouse gas, diesel is slightly better than gasoline, and coal is almost twice as polluting as natural gas. Note that the numbers for automobile fuels are not directly indicative of how much pollution is emitted per mile of driving. Electricity is by far the most polluting energy source. At present, most of our electricity is generated from coal, the most polluting fossil fuel we have, and power plant turbines are only one-third efficient.

Source: EIA, "Emissions of Greenhouse Gases Report," December 2008, www.eia.doe.gov/oiaf/1605/archive/gg08rpt/index.html.

in coal into electricity. Not only does coal produce more pollution than other fuels, more of it is needed to generate electricity than if it were used directly for heating. The overall effect is that coal produces a lot of pollution for the amount of useful electrical energy it produces.

Conventional wisdom says that electricity is exceptionally clean and pollution-free. Indeed, this is one of the points in favor of electric cars. However, generating electricity currently contributes a third of all greenhouse gas emissions, half again as much as gasoline and diesel use combined. Although using electricity for vehicle propulsion is very clean, generating the electricity is very dirty. If one wants to go after the leading single source of greenhouse gases, one should go after coal-fired electric power plants rather than gasoline-

TABLE 1.4 *Pollution by Energy Content of Fossil Fuels*

Pollutant	Pollution (lb of CO_{2e} per billion Btu)		
	Natural gas	Oil	Coal
Carbon dioxide	117,000	164,000	208,000
Carbon monoxide	40	33	208
Nitrogen oxides	92	448	457
Sulfur dioxide	1	1,122	2,591
Particulates	7	84	2,744
Mercury	0.0	0.007	0.016

Source: Natural Gas Supply Association, "Natural Gas and the Environment; Fossil Fuel Emission Levels," www.naturalgas.org/environment/naturalgas.asp.

burning automobiles. Of course, that would do nothing for reducing demand for gasoline and our dependence on foreign oil.

Finally, I should mention pollution other than greenhouse gases and CO_2. Table 1.4 shows the pollution, including the greenhouse gas carbon dioxide, emitted by fossil fuels. The first thing to notice is that amounts of pollutants are much smaller than amounts of CO_2, 100 to 1 or 1,000 to 1. The most important point to note is how much more polluting coal is than natural gas (oil is between the two). Natural gas is, indeed, much cleaner than either oil or coal.

Petroleum

Petroleum—oil, crude, black gold—is the main concern for several reasons. Cars and trucks are almost all powered by gasoline and to a lesser degree by diesel, and both come from petroleum. The United States imports over half of its oil, and gasoline and diesel are major sources of pollution and greenhouse gases.

People have known about and used petroleum (from the Latin: *petr*, or *rock*, and *oleum*, or *oil*, i.e., *rock oil*) for centuries.[9] In ancient times, Babylon used asphalt for construction, and Persia used petroleum for medicines and lighting. The Chinese drilled wells to obtain petroleum to use as a fuel to evaporate water and obtain salt. Around the ninth century, Arabs and Persians used it for lighting and military applications.

One major line in the development of the petroleum industry involved lighting. People discovered how to refine kerosene from coal around 1850 and from petroleum shortly thereafter. Almost simultaneously, commercial development of oil started in Romania in 1857, followed by Canada in 1858, and the United States in 1859, when oil was discovered at Titusville, Pennsylvania. Petroleum had become abundantly available, and the demand for kerosene and oil for lamps drove early development. Among other things, these discoveries decreased the demand for whale oil and probably saved several whale species from extinction.

A second line of development started in the mid-nineteenth century, when Western Europe started synthesizing chemicals that could substitute for natural products. Then, in World War I, the British learned how to extract benzene and toluene from oil. These are important feedstocks for the petrochemical industry. Since then, the petrochemical industry has grown to such an extent that modern civilization depends on the synthesized materials. The petrochemical industry provides plastics, soaps, detergents, solvents, paints, medicines, fertilizer, pesticides, explosives, synthetic fibers like nylon and polyester, synthetic rubber, flooring, and insulating materials.

A third line of development was automobile fuel. While early automobiles were propelled by pedal power, hydrogen and oxygen internal combustion engines, steam, and even electricity, the discovery of petroleum and the development of inexpensive gasoline and diesel fuels spurred demand for vehicles and fuels. Gasoline and diesel became the major products of crude oil.

Oil Refining

Petroleum contains hydrocarbons and other organic compounds. Hydrocarbons, which are molecules consisting of hydrogen and carbon, become useful products. The other organics, such as nitrogen, oxygen, and sulfur compounds, are pollutants, although they may be useful for some products. Different hydrocarbon molecules have different useful properties, different numbers of carbon atoms per molecule, and different evaporation temperatures. The modern refining process takes advantage of the different evaporation temperatures to separate out the different "fractions" and isolate the desired hydrocarbon molecules.[10] The fundamental process in refining a barrel of crude is selectively and sequentially boiling off the different hydrocarbons. First come various gases, such as liquefied petroleum gas (LPG), then gasoline, then diesel, then several different fuel oils (heating oil, fuel oil, jet fuel, etc.), and finally other products that are primarily feedstock for the petrochemical industry, which gives us such things as plastics and pharmaceuticals. This is not the end of the processing. The gasoline that comes out of the distillation process, "straight run" gasoline, is of relatively poor quality. Further "cracking" (breaking up longer hydrocarbon chains into shorter chains and straight chains into branched chains) is required to produce high-quality gasoline. By the end of the refining process, each 42-gallon barrel of oil becomes a little over 44 gallons of product. This might seem strange, but normal refining processes involve cracking long-hydrocarbon-chain molecules into shorter-chain molecules, making more but less-dense molecules. Figure 1.5 shows the breakdown of refining in the United States.

Figure 1.5 illustrates a very important point. Petroleum is a mixture of numerous different hydrocarbon molecules, and the refining process simply separates the different components. Refining does not convert one hydrocarbon molecule into another. This has two implications for us. First, two-thirds of a barrel of oil is gasoline or diesel, and the remaining one-third becomes other products that are essential to modern society, such as fuel oil, lubricating oil, plastics, and pharmaceuticals. Eliminating demand for gasoline and diesel would not reduce demand for oil because we would still need the other products. Gasoline and the other products are not interchangeable. You cannot convert gasoline into petrochemi-

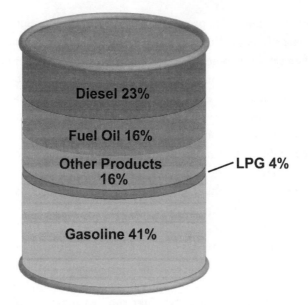

Diesel 23%

Fuel Oil 16%

Other Products 16%

LPG 4%

Gasoline 41%

FIGURE 1.5 Products from US Petroleum Refineries, 2008. Each 42-gallon barrel of crude oil becomes 44 gallons of product during the refining process. The percentages of various products depend on the mix of crude oils and adjustments to the refining process. Each barrel of oil yields about twice as much gasoline as diesel fuel. About 16% of each barrel becomes nonfuel products such as plastics and pharmaceuticals. Note that the percentages change slightly from year to year.

Source: EIA, "Oil: Crude and Petroleum Products Explained," http://tonto.eia.doe.gov/energy explained/index.cfm?page=oil_home.

cal feedstock, for example. To eliminate demand for oil, we need to find replacements for fuel oil, plastics, and fertilizer, among other things. Second, the amounts of gasoline and diesel in petroleum are relatively fixed. Conventional wisdom extols the desirability of diesel engines over gasoline engines because diesel engines are more efficient. Shifting to diesel results in consumption of fewer gallons of automobile fuel, but because a barrel of crude oil contains less diesel fuel than gasoline, demand for petroleum might actually increase as a result of a general shift to diesel fuel. I return to this subject in chapter 3.

Not all petroleum is equal. Crude is categorized by where it comes from (e.g., West Texas), its density (light or heavy), and its sulfur content ("sweet crude" does not contain sulfur; "sour crude" does). Location is important because it affects transportation costs and control by foreign governments. Density is important because light crude is easier to transport and pump and contains more hydrocarbons (particularly gasoline) than heavy crude does. Sweetness is important because sour crude presents more severe environmental issues and requires more processing to minimize pollutants. At a more detailed level, the amount of gasoline (as well as other fractions) in a barrel of crude varies with the country of origin. With basic refinery processing, a barrel of sweet light crude produces about 30% gasoline. By contrast, a barrel of heavy sour crude yields about 14% gasoline, but Venezuelan crude produces only about 5% gasoline. Venezuelan crude is good for fuel oil and heating oil but is not a good source of gasoline. Perhaps this is why Hugo Chavez, through US-incorporated but Venezuelan-owned CITGO Corporation, can give away heating oil to the US poor. US refineries are yielding about 41% gasoline per barrel because of the mix of crudes they process and the extensive postdistillation processing (cracking, reforming, etc.), which can greatly improve yields but involves increased complexity and cost. Refineries in Europe operate with a different mix of input crudes and different processing methods because Europe requires more diesel fuel than the United States does. European refineries produce about the same percentage of petrochemical feedstocks and LPG as do US refineries, but the Europeans produce more fuel oil and diesel and less gasoline than do the Americans. While about 70% of the combined US and European production of gasoline comes from the United States, Europe accounts for about 60% of the combined production of diesel.

The single most important reason for the differences between outputs from US and European refineries is the mix of crude oils going into the refineries. We need an active international trade in *crude oil* to provide the desired mix of petroleum for our refineries, which allows them to match the range of products to the range of US demand. We need active trade in *petroleum products* to compensate for remaining mismatch between refinery output and demand. We will need to trade and be dependent on foreign oil to some degree for as long as we need petroleum products.

Production, Consumption, and Imports

US annual consumption of petroleum is 7.1 billion barrels. A primary concern is that US imports of crude amount to 66% of consumption. How did we get into the position of importing so much oil?

US demand for oil has steadily increased for the past hundred years (fig. 1.6). The first line in figure 1.6 shows US crude-oil production. Production of domestic crude oil grew smoothly until 1971 and has been declining ever since. The second line shows imports. The United States has had to import more oil each year since 1971 to balance increasing demand with decreasing domestic production. Imports were insignificant before the 1940s and increased slowly and slightly in the 1940s, 1950s, and 1960s. Imports started increasing markedly around 1971, plummeted around 1977, and started increasing again in 1985. The 1973 Arab oil embargo in retaliation against the United States for supplying Israel during the Yom Kippur war and the 1979 oil crisis are partially responsible for the volatility around 1980. Note the sharp increase in imports in the decade before the embargo. One has to wonder if the embargo had more to do with the unreasonable increase in US demand for foreign oil than with our support for Israel. In any case, the growth in US imports returned to a more moderate rate after the embargo. The third line in the figure shows demand. This is actually what the EIA calls "Refinery and Blender Net Input of Crude Oil" (i.e., the crude oil and refined product that goes into the US production process). Bear in mind that this includes both domestically produced and imported crude. Data are not available prior to 1980, but a little thought shows that this is unimportant. Prior to 1971, imports were quite small, and domestic production of crude was a good measure of total consumption. As imports grew, domestic production became less indicative of total consumption, and refinery input, the demand graph, became a better measure. From 1981 on, the data for refinery input have been a good measure of total consumption.

Almost half of the oil we import comes from the twelve member nations of the Organization of Petroleum Exporting Countries (OPEC): Algeria, Angola, Ecuador, Iran, Iraq, Kuwait, Libya, Nigeria, Qatar, Saudi Arabia, United Arab Emirates, and Venezuela. We also import moderate amounts of oil from Canada, Mexico, the Virgin Islands, and Russia (fig. 1.7) and smaller amounts of oil from seventy-eight other non-OPEC countries. Conventional wisdom seems to be that well over half of our oil comes from the Mid-

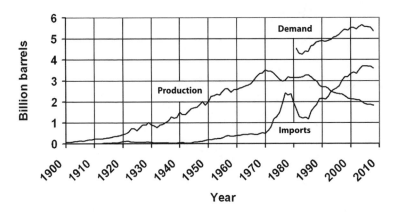

FIGURE 1.6   US Oil Production, Demand, and Imports. Demand for oil has increased steadily for the past hundred years. Domestic production kept pace with demand until about 1970, with almost no imports of foreign oil. Domestic production has declined steadily since 1970. Continued steady growth in demand has required steady growth in foreign imports.

Sources: Production: EIA, Petroleum Navigator, "US Field Production of Crude Oil," http://tonto.eia.doe.gov/dnav/pet/hist/mcrfpus2a.htm. Imports: EIA, Petroleum Navigator, "Annual US Imports of Crude Oil," http://tonto.eia.doe.gov/dnav/pet/hist/mcrimus 1a.htm. Consumption: EIA, Petroleum Navigator, "Annual US Product Supplied of Crude Oil and Petroleum Products," http://tonto.eia.doe.gov/dnav/pet/hist/LeafHandler.ashx?n= PET&s=MTTUPUS1&f=A.

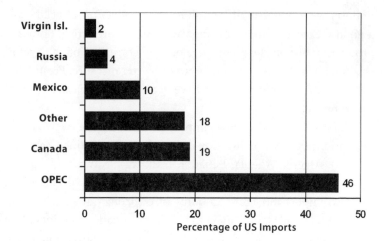

FIGURE 1.7 US Oil Imports by Country, 2008. The United States imports 66% of its oil. About half of our imported oil (31% of US consumption) comes from OPEC.

Source: Percentages calculated from EIA, Petroleum Navigator, "US Imports by Country of Origin," http://tonto.eia.doe.gov/dnav/pet/pet_move_impcus_a2_nus_ep00_im0_mbbl _a.htm.

dle East, particularly Saudi Arabia. This just is not true. It is true that 66% of our oil is imported, but less than half of that comes from OPEC, and not all of that comes from the Middle East. In terms of total oil consumption, not just imports, 31% comes from OPEC, 12% comes from the Middle East, and less than 8% comes from Saudi Arabia.

Reserves

The next question is, how long will oil last? Perhaps more to the point, how much oil is there? That is, how large are the oil reserves? There is great discussion and disagreement over this question, and we will get to some of the issues shortly.

Domestic proved reserves increased steadily until 1960 (fig. 1.8), followed by steady decline since then. Discovery of the huge Prudhoe Bay field in Alaska in 1971 gave a short-lived boost to

FIGURE 1.8   Proved US Oil Reserves by Year. Spurred by exploration and oilfield development, US proved reserves of oil grew steadily until the 1960s and then leveled off. The discovery of the huge Prudhoe Bay deposit in 1970 produced a large but temporary jump in reserves. Since then, domestic proved reserves have steadily declined. The latest data for 2008 show proved reserves at 19.1 billion barrels, enough for seven years at our current rates of consumption, imports, and domestic production.

Source: EIA, Petroleum Navigator, "Annual US Crude Oil Proved Reserves," http://tonto.eia .doe.gov/dnav/pet/hist/LeafHandler.ashx?n=PET&s=RCRR01NUS_1&f=A.

US reserves. With the exception of Prudhoe Bay, the US petroleum industry is not currently discovering or developing new deposits of oil, and there is no expectation of discovering new deposits of oil similar to Prudhoe Bay. Reserves and domestic production are both decreasing. US reserves in 2008 were 19.1 billion barrels. The United States consumes 7.1 billion barrels of crude each year, 2.6 billion barrels of which come from domestic production. Domestic reserves will last about seven years at the current rate of production. With increasing demand and decreasing domestic production, the United States will import increasing amounts of oil in coming years. We could stabilize imports by increasing production, but that would hasten the depletion of our reserves.

There are several different types of reserves in the energy literature, such as "proved," "measured," "indicated," "demonstrated," and "prospective." Each term means something different, and the numbers of barrels of oil in each type of reserve differ substantially. Making sense of the discussion of reserves is a challenge. The data I have presented reflect *proved reserves*, the most conservative estimate. The location, quantity, and grade of the energy source in these reserves are well established and uncontroversial. The EIA provides the following definition of proved reserves:

> Proved reserves of crude oil . . . are the estimated quantities of all liquids defined as crude oil, which geological and engineering *data demonstrate with reasonable certainty* to be recoverable in future years from *known reservoirs* under *existing economic and operating conditions*. . . . Reservoirs are considered proved if economic producibility is supported by *actual production or conclusive formation test* (drill stem or wire line), or if economic producibility is supported by core analyses and/ or electric or other log interpretations. The area of an oil reservoir considered proved includes: (1) that portion delineated by drilling and defined by gas—oil and/or gas—water contacts, if any; and (2) the immediately adjoining portions not yet drilled, but which can be reasonably judged as economically productive on the basis of available geological and engineering data. In the absence of information on fluid contacts, the lowest known structural occurrence of hydrocarbons is considered to be the lower proved limit of the reservoir. . . . Reserves of crude oil which can be produced economically through application of improved recovery techniques (such as fluid injection) are included in the "proved" classification when *successful testing by a pilot project*, or the operation of an installed program in the reservoir, provides support for the engineering analysis on which the project or program was based.[11]

Succinctly, with regard to proved reserves, we know the crude oil is there, and we know how much is there because we have seen it and touched it and measured it. One can be confident of proved reserves.

There may be more oil if we consider *undiscovered technically recoverable reserves* (UTRR). The US Department of the Interior Minerals Management Service (MMS) estimate of UTRR on the outer continental shelf in 2006 was 85.9 billion barrels (mean value).[12] The United States Geological Survey (USGS) estimate of onshore UTRR in 2007 was 42.1 billion barrels.[13] Together these amount to 128 billion barrels of crude, six times the proved reserves. These estimates span a range of probable values. The numbers given here are mean values. That is, there is a 50% probability that we will actually find 128 billion barrels, but there is a 95% probability that we will find at least 100 billion barrels and a 5% probability of finding 170 billion barrels or more.

I am uneasy with the rosy UTRR estimates, and not just because the estimates are probabilistic and therefore uncertain. My uneasiness stems from the definition of UTRR. The MMS assessment report defines UTRR as "Oil and Gas that may be produced as a consequence of natural pressure, artificial lift, pressure maintenance, or other secondary recovery methods, but *without any consideration of economic viability*. They are primarily located *outside of known fields*."[14]

The Bakken Formation exemplifies my concern with prospective reserves. The Bakken Formation is a 200,000-square-mile area extending across Montana, North Dakota, and Saskatchewan containing large oil and gas reserves. Although there have been several published estimates of the volumes of oil and natural gas in the formation, there is no agreement on the actual volume of remaining resources. Estimates change from time to time as data, methodology, and assumptions change. The current USGS estimate is 3.65 billion barrels of oil in UTRR of the Bakken Formation.[15] This is a twenty-five-fold increase over the previous estimate of 151 million barrels made in 1995. The magnitude of the change makes me suspicious of the estimation process.

We *know* that the United States has oil reserves totaling 19.1 billion barrels. We *think* there might be an additional 128 billion barrels, for a total of 149 billion barrels. In that case, our reserves would last fifty-seven years rather than seven at current rates of consumption, production, and importation. Regardless of which prediction turns out to be true, we are going to have severe problems with supply of oil within the lifetime of children being born now. Where the future of civilization as we know it is concerned, I would prefer to stick to proved reserves.

Proved oil reserves worldwide (1,332 billion barrels) and rate of consumption (31.2 billion barrels per year) are such that the world will not run out of oil for forty-three years if consumption remains at the current level. Oil reserves are concentrated in the Middle East and OPEC nations, as shown in figure 1.9. OPEC controls three-quarters of world oil reserves, and most of that is in the Middle East. The United States, and much of the world, will be dependent on OPEC for energy for years to come.

Bottom Line

The United States will run out of domestic proved oil reserves in about seven years if we continue the current rates of consumption, production, and importation. Unfortunately, demand is growing and production is decreasing. Imports will probably have to increase, but this is just what we want to avoid. We can stabilize imports and minimize foreign control by increasing production, which means developing new wells and tapping our reserves, but this will hasten depletion of our reserves. How long we can postpone depleting our reserves depends on how successful we are in developing undiscovered oil fields. Even rosy predictions indicate trouble within the lifetime of children being born today. We can also postpone it by reducing oil consumption. One way of reducing demand for oil is reducing demand for gasoline and diesel. I discuss alternative-vehicle technology in chapter 3.

Becoming independent of foreign oil is a tall order. Conventional wisdom is that by reducing gasoline consumption, we will reduce demand for foreign oil. I disagree. Even if we were to elim-

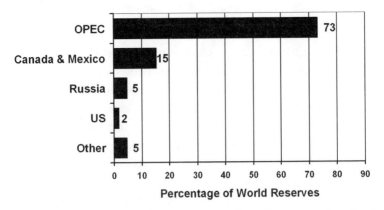

FIGURE 1.9 Proved World Oil Reserves. Proved world reserves of crude oil stand at 1,332 billion barrels, enough for about forty years at the current world consumption rate of 31.2 billion barrels per year. OPEC, with 73% of world reserves, and the Middle East, with 59% of reserves, dominate the world's oil supply.

Source: Percentages calculated from data at EIA, "International Energy Statistics; Crude Oil Proved Reserves," http://tonto.eia.doe.gov/cfapps/ipdbproject/IEDIndex3.cfm?tid=5&pid=57&aid=6.

inate gasoline consumption, we would still need other petroleum products, such as fuel oils, pharmaceuticals, and petrochemical feedstock. Unless we find replacements for these or strongly reduce consumption, reducing gasoline consumption will have a marginal effect on demand for foreign crude. Second, because the percentages of various products from refining change with the mix of incoming crude, we need an active international trade in the various grades of crude in order to have efficient refining for as long as we need oil.

Natural Gas

Natural gas is available domestically, it burns cleanly, and transporting it is easy. Moreover, it is a vehicle fuel (see chapter 3). Natural gas has a lot to offer.

Transportation and Storage

Transporting natural gas via pipeline under pressure is straightforward. Transportation domestically and between the United States and Mexico and Canada is by pressurized pipeline. There are over 210 pipeline systems in the United States, with over 305,000 miles of pipes and fourteen hundred compressor stations for maintaining pressure in the pipelines. There are also around four hundred underground storage fields. Common practice is to smooth demand on the pipeline network by storing natural gas during April through October and extracting it during the November to March heating season. Pipeline storage capacity is 1.6 to 3.6 Tcf, reasonably well matched to the annual home heating demand. Pipelines make a good transportation system. Gas flows readily, capacity is easily increased by raising the pressure in the pipeline (within reason of course), and the energy to pump and compress the natural gas in the pipeline comes from the gas itself. Operators siphon off about 8% of the gas flow to operate the pipeline, so there is a slight energy loss inherent in pipeline transportation.

Because of the low energy content of natural gas at atmospheric pressure (1,031 Btu per cubic foot), long-distance and transoceanic transport requires another approach. In terms of energy content, it would take over 900 gallons of natural gas at atmospheric pressure to provide the energy contained in a single gallon of gasoline. Natural gas at atmospheric pressure is an inefficient medium for storage and transportation because of the large volume of gas needed. There are two ways to address this problem. The first is to transport the resource in the form of liquefied natural gas (LNG), which is natural gas cooled to –260°F. Cooling the gas condenses it about 600:1 by volume and turns it into a liquid. LNG has slightly less energy per unit volume than gasoline, but it is an efficient form for storage. It is more efficient to transport natural gas as LNG in large, cooled tanks, which are essentially giant refrigerated thermos bottles. Transoceanic shipping and trucking to places not served by pipelines use LNG. LNG is warmed and converted back to gas at the destination. As in pipeline operations, common practice is to siphon off some of the gas as the source of energy for cooling the gas initially, keeping the tanks cold during transport, and warming the gas at the destination. About 15% of the gas is lost to cooling during a typical transoceanic voyage.

The other method relies on compressed natural gas (CNG). The gas is compressed to about 3,600 psi (pounds per square inch) at normal ambient temperature. The energy content per volume of CNG is only about a quarter of that of LNG, but insulated storage tanks and active cooling are not necessary. The more complicated to produce and expensive LNG is preferred for long-distance transportation because of the higher energy per volume, and CNG is preferred for short-distance transportation and long-term storage because active cooling is not required. Still, CNG requires strong, heavy tanks to contain gas at 3,600 psi. I return to this matter in chapter 3 in the discussion of natural gas vehicles.

Production, Consumption, and Imports

Figure 1.10 shows how we used natural gas in the United States in 2008. Most of the natural gas went to domestic use in heating, cooling, and cooking and to generating electricity. The "Plant & Pipeline" category in the figure includes gas consumed in plant and pipeline activities and a very small amount used for vehicle fuel.

US consumption and production of natural gas have been increasing slowly since the late 1940s (fig. 1.11), with consumption growing slightly faster than production. Consumption is currently 23.2 Tcf and domestic production a little under 20 Tcf. To offset the difference, the United States imports 3.98 Tcf annually, almost all of it from Canada, but some from Mexico, and some from outside North America, mainly from Trinidad (fig. 1.12). Despite claims of huge domestic supply, the United States imports 16% of its natural gas, far less than the 66% of our petroleum that we import, but still significant.

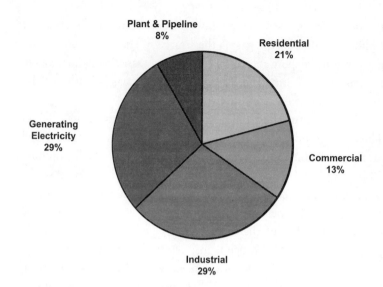

Plant & Pipeline
8%

Residential
21%

Generating
Electricity
29%

Commercial
13%

Industrial
29%

FIGURE 1.10 US Natural Gas Consumption by Sector, 2008. Most natural gas goes to heating. Powering and maintaining the pipeline network accounts for 8%. Almost none is used in transportation. About 29% is used to generate electricity, compared with 100% of uranium, 1% of oil, and 91% of coal.

Source: Percentages calculated from data at EIA, Natural Gas Navigator, "Natural Gas Consumption by End Use," http://tonto.eia.doe.gov/dnav/ng/ng_cons_sum_dcu_nus_a.htm.

Reserves

Conventional wisdom is that we have plenty of natural gas. Is the conventional wisdom correct? How much natural gas do we have in the United States? Can we continue to be gas-independent of the rest of the world? Well, not entirely. Proved US reserves amount to 237.7 Tcf, which will last about twelve years at our current rate of production. The United States will have to increase imports if domestic production is not increased. Indeed, this is already happening, as evidenced by the number of new LNG terminals being planned or constructed. We will soon be in the same position of dependence on foreign sources of natural gas as we now are on foreign sources of petroleum. Of course, as exploration continues

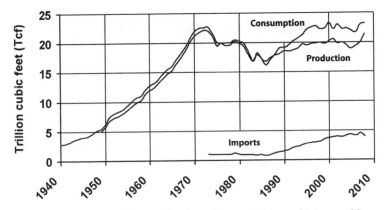

FIGURE 1.11   US Natural Gas Production, Consumption, and Imports. Natural gas production and consumption grew steadily until about 1971, when the discovery of the Prudhoe Bay oil deposit boosted oil supply. Production and consumption decreased for over a decade and then resumed a steady increase about the time that oil imports started increasing after the temporary drop brought on by the oil embargo. Current consumption is 23 Tcf, of which 3.98 Tcf (16%) is imported, predominantly from Canada.

Sources: Consumption: EIA, Natural Gas Navigator, "Annual US Natural Gas Total Consumption," http://tonto.eia.doe.gov/dnav/ng/hist/n9140us2a.htm. Production: EIA, Natural Gas Navigator, "Annual US Natural Gas Marketed Production," http://tonto.eia.doe.gov/dnav/ng/hist/n9050us2a.htm. Imports: EIA, Natural Gas Navigator, "Annual US Natural Gas Imports," http://tonto.eia.doe.gov/dnav/ng/hist/n9100us2a.htm.

and prices of fuel rise, more natural gas will become economically available. On the other hand, extensive conversion from gasoline vehicles to natural gas vehicles will increase natural gas demand and severely exacerbate the supply situation.

World consumption of natural gas is currently 105.5 Tcf annually. Comparing this consumption rate with proved world reserves of 6,254 Tcf indicates that the world will not run out of natural gas for fifty-nine years if consumption remains at the current level. Compared to world reserves of oil, natural gas reserves in the Middle East are smaller, and reserves in the Western Hemisphere and Russia are larger (fig. 1.13).

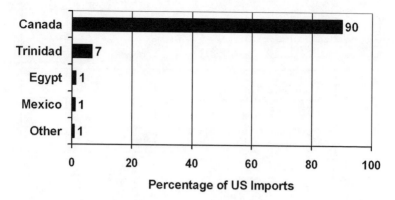

FIGURE 1.12 US Natural Gas Imports by Country, 2008. US imports of natural gas amount to 3.98 Tcf per year, 16% of consumption, almost all of it from Canada.

Source: Percentages calculated from data at EIA, Natural Gas Navigator, "US Natural Gas Imports by Country," http://tonto.eia.doe.gov/dnav/ng/ng_move_impc_s1_a.htm.

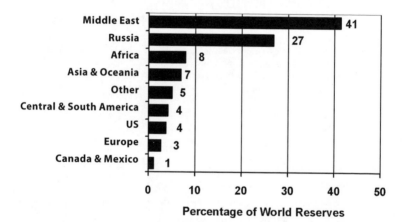

FIGURE 1.13 Proved World Natural Gas Reserves, 2009. Total proved world natural gas reserves are 6,254 Tcf, enough for about forty years at current and projected rates of consumption. While the Middle East holds about 41% of the world reserves of natural gas (compared with almost 60% of the oil), it will still be a major player in energy as we depend more heavily on natural gas. The dominant suppliers will be in the Middle East and Russia.

Source: Percentages calculated from data at EIA, "International Natural Gas Reserves and Resources," www.eia.doe.gov/pub/international/iealf/naturalgasreserves.xls.

Bottom Line

Natural gas is a clean fuel used primarily for space heating and cooling, cooking, and generating electricity. US production and consumption are relatively constant, and US proved reserves should continue to be adequate for the next ten years with constant levels of consumption, production, and imports. Natural gas will last longer than oil at current rates of production and consumption, but the United States will have to increase imports relatively soon. Fortunately, world consumption and reserves are such that the worldwide supply is secure for a longer time, about sixty years. This situation could change drastically if demand for natural gas increases sharply as a result of widespread use of natural gas vehicles.

The United States currently imports 16% of its natural gas. Trends in production and consumption strongly indicate that we will have to import a larger percentage of our natural gas in the future despite the commonplace claims of "plentiful" domestic natural gas. The number of new LNG terminals in planning or under construction convincingly point to the likelihood of increased imports of natural gas. In addition to having to increase foreign imports of natural gas to satisfy domestic demand, a significant percentage of the gas will be lost because of the need to cool the gas during transportation and reheat it at its destination, effectively increasing consumption.

Coal

Coal is the most abundant and widely distributed fossil fuel worldwide. Coal is primarily carbon with varying amounts of sulfur, nitrogen, oxygen, and hydrogen. Coal is the primary source of electricity worldwide and is the primary source of CO_2 emissions and pollution worldwide. Mining coal is extremely damaging to the environment and to the health and safety of miners.

There are several types of coal with differing energy and carbon content: peat, lignite, subbituminous, bituminous, anthracite, and graphite.[16] Peat and lignite are not major players. Graphite does not burn very well but makes great pencils and powdered lubricant. For our purposes, bituminous coal and anthracite are the most important. Anthracite is the hardest of the lot and has the most energy per pound and the least pollutants. It is the main coal used for residential and commercial space heating. Bituminous coal is used mainly for electric power generation and combined heat and power in industry and as a source of aromatic hydrocarbons for the chemical industry.

Production, Consumption, and Imports

US consumption of coal in 2007 was 1.1 billion short tons.[17] About 93% is subbituminous or bituminous, and less than 1% is anthracite. Ninety-one percent goes to electric power utilities, but industrial production of electricity raises the percentage of coal that goes to generating electricity. Almost all coal, 91% of it in the United States, is used to generate electricity.

US production of coal in 2008 was 1.17 billion short tons, slightly greater than consumption (1.12 billion short tons). The US exports some coal and imports some. Net trade comes to 47.3 million short tons exported. This is a mere 4% of production. For all practical purposes, export and import rates may be ignored at the current time.

Reserves

Coal reserves present a situation vastly different from those of oil. US recoverable coal reserve is 271 billion short tons,[18] enough for 226 years at the current rate of production. World reserves are 1,001 billion short tons, enough to last 143 years at current rates of consumption. The United States will still have coal after the rest of the world runs out. Indeed, 75% of the world's coal is in the United States, the United Kingdom, Russia, China, and India (fig. 1.14).

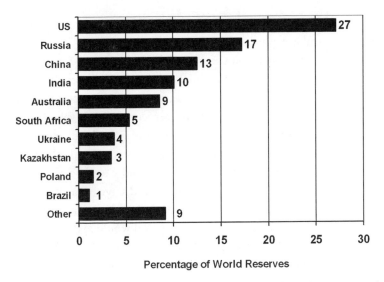

Percentage of World Reserves

FIGURE 1.14 Proved World Coal Reserves, 2006. Total world reserves of coal amount to 1,001 billion short tons, of which 272 billion short tons are in the United States. Coal is the one fossil fuel the United States has in abundance. Should the world shift from oil to coal, dominance would shift away from OPEC and the Middle East toward the United States, Russia, China, India, and Australia, which together hold three-quarters of the world's coal.

Source: Data from *BP Statistical Review of World Energy June 2007*. The data are also found in readable format at Wikipedia, "Coal," table "Proved Recoverable Coal Reserves, at End-2006," http://en.wikipedia.org/wiki/Coal. Note that the data have been converted from metric ton to short tons.

The Middle East is devoid of coal. If current rates of production and consumption are maintained, the United States will eventually be a supplier of coal to the rest of the world.

Bottom Line

Domestic coal reserves will last a couple of hundred years at current rates of production and consumption. Worldwide reserves will last a shorter time. That is the good news. The bad news is that almost all coal consumed in the United States goes to generating electricity, with the result that coal is responsible for huge amounts of pollution and greenhouse gases.

The role coal plays in the national energy strategy in coming decades depends on several conflicting factors. First, coal is a hydrocarbon fossil fuel that, in contrast to oil and natural gas, is abundant in the United States. However, coal is not a vehicle fuel, so it cannot directly replace oil or natural gas. As discussed in chapter 4, there is ongoing research into converting coal into liquid vehicle fuel. If these efforts are successful and we can get gasoline from coal without increasing pollution and environmental damage, then we are in a very good position. Demand for foreign oil would then decrease. Demand for foreign oil would also decrease if there were a paradigm shift away from internal combustion engines to electric cars. Coal then becomes the primary source of vehicle fuel because of its preeminent role in generating electricity. However, the inherent dirtiness of coal is a significant impediment. As discussed in chapter 4, there is ongoing research into making coal "clean." If this is successful, it will significantly improve our energy situation.

Uranium

Like coal, uranium is not an automotive fuel, but it is a source of electricity, so it belongs in any discussion of driving green. As a means of generating electricity, uranium (which fuels nuclear power), is unique among the energy sources we have been discussing in that it is not a fossil fuel and produces no conventional pollution or greenhouse gases. On the down side, there is strong opposition to nuclear power in the United States because of safety and security concerns.

Simply stated, a nuclear reactor brings about fission of uranium atoms to heat water that drives a turbine to generate electricity. The reactor contains uranium fuel rods, which last three to five years, after which time they become "spent" and are replaced. The reactor generates heat in a controlled chain reaction. Water, or some other coolant, circulates through the reactor and transfers heat from the reactor to a boiler. The hot water or steam from the boiler then drives a turbine, which generates electricity.

Electricity Production

There are 65 nuclear power plants in the United States with 104 individual reactors (fig. 1.15). The first full-power operating license was granted in 1957. The last construction permit was issued in 1979, and no new reactor has started operations since 1997. These plants provide 100,000 megawatts (MW) total capacity, and they generated 806 billion kWh of electrical energy in 2008.[19] This was 20% of US consumption (coal, natural gas, and oil provide the rest).

Worldwide, there are 436 nuclear power reactors including the 104 in the United States. There are operating reactors in 31

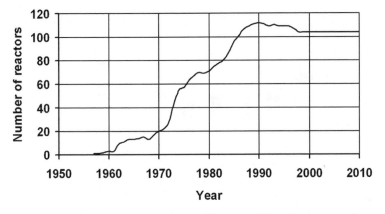

FIGURE 1.15 US Nuclear Power Reactors. There are 65 nuclear power plants in the United States with 104 individual reactors. The last construction permit was issued in 1979, and no new reactor has started operations since 1997. Several reactors have been deactivated since 1990, when the number of active reactors peaked at 112.

Source: EIA, "Annual Energy Review, 2009," table 9.1, "Nuclear Generating Units, 1955–2009," www.eia.doe.gov/emeu/aer/pdf/pages/sec9_3.pdf.

countries, with plans or proposals for reactors in 15 more. Nuclear power provides 2% of the world's energy needs and 15% of the world's electricity. While the United States uses more uranium and produces more electricity from nuclear power than any other country (fig. 1.16), other countries are more dependent on nuclear power. The United States generates 21% of its electricity using nuclear power, the European Union 30%, and France 80%. Opposition to nuclear power notwithstanding, nuclear power for civilian electricity is commonplace.

Production, Consumption, and Imports

Uranium ore contains several different oxides of uranium, and different ores from different locations contain different oxides and in varying concentrations. That is, there is wide variation in ura-

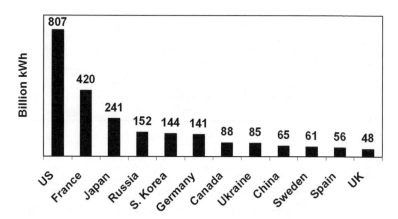

FIGURE 1.16   Nuclear Power Production of Electricity by Country, 2007. Worldwide, thirty-one countries generate at least some of their electricity from nuclear power. Total generation of nuclear power electricity was 2,595 billion kWh in 2007. The twelve largest producers, shown here, generate 89% of the world's nuclear power electricity. The United States produces 31%. The United States has more reactors than any other country and produces more electricity by nuclear power than any other country.

Source: Data from EIA, "Table 2.7 World Net Nuclear Electric Power Generation, 1980–2007," www.eia.doe.gov/pub/international/iealf/table27.xls.

nium ores. Fundamental processing consists of grinding, milling, and chemical extraction of the uranium oxides, resulting in "yellowcake." Modern yellowcake is typically 70% to 90% triuranium octaoxide (U_3O_8) by weight, the remainder being made up of other oxides, such as uranium dioxide (UO_2) and uranium trioxide (UO_3). Discussions of uranium consumption actually refer to consumption of U_3O_8 or its equivalent.

World consumption of uranium is 150 million pounds a year.[20] World production of uranium in 2008 was 96.7 million pounds.[21] This means that the world's production from uranium mines covers less than three-quarters of world demand. This situation is not as dire as it might sound at first. The deficit is made up by withdrawing material from uranium stockpiles, converting weapon materials to commercial use, reprocessing spent reactor fuel rods, and re-enriching tailings left over from the initial enrichment processing. While some of these measures, such as withdrawing from stockpiles and converting nuclear weapons, will soon end, others will continue. As prices increase and techniques improve, reprocessing tailings will become more productive. Reprocessing spent fuel rods also expands the supply of uranium. Together, reenriching tailings and reprocessing fuel rods adds 12% to 20% to the uranium supply. Moreover, mines traditionally have been operating at three-fourths capacity, so productivity could be increased.

The picture becomes very interesting when we look at indigenous supplies of uranium. Indeed, 60% of production comes from three countries: Canada, Australia, and Kazakhstan. This means that a very large number of countries depend on a few suppliers.

US consumption is 51.3 million pounds per year (uranium loaded into reactors). Domestic production provides 4 million pounds, and net imports provide 39.9 million pounds. As in the world at large, the deficit is made up with material from stockpiles and converting weapons. The United States does not reprocess reactor fuel rods, so that source is not available. Overall, the United States imports about 80% of its uranium.

Reserves

World reserves are estimated to be 12 billion pounds.[22] With world consumption running at 150 million pounds per year, reserves should last eighty years. However, this estimate does not consider the techniques currently used to make up the deficit between production and consumption. Withdrawals from stockpiles, converting nuclear weapons to civilian power needs, reprocessing tailings, and reprocessing spent reactor fuel rods will stretch reserves between 12% and 20%. That is, world reserves should last almost a hundred years with no growth in demand.

Correspondingly, US reserves of uranium are 890 million pounds, which will support the current production rate for over two hundred years as long as the high rate of imports is maintained. US reserves would support current consumption for about seventeen years if foreign imports were curtailed.

The last thing to look at is where the reserves are located (fig. 1.17). Australia, Kazakhstan, Russia, South Africa, Canada, and the US hold 75% of world uranium reserves; the United States has slightly less than 5%.

Potential

Nuclear power has the potential for being a limitless source of energy. Fast breeder reactor technology extends the promise of producing almost as much nuclear fuel as it consumes, thereby extending reserves hundreds or thousands of years. Feasibility demonstrations of extracting uranium from seawater, an almost limitless reserve, also show promise. There are concerns with both of these possibilities, but if they pan out, the world supply of electricity will be secure for a long time.

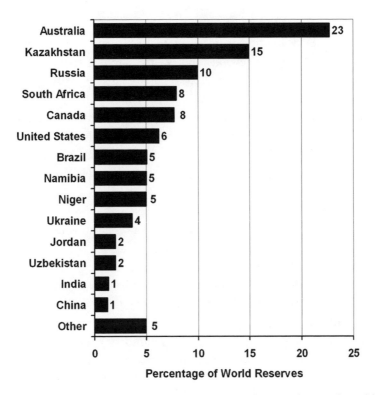

FIGURE 1.17 Proved World Uranium Reserves, 2007. Total proved world reserves of uranium were estimated to be 12,000 million pounds (5.5 MMt) in 2007. This will last eighty years at current rates of consumption. There is almost no uranium in the Middle East.

Sources: World Nuclear Association, "World Uranium Mining," table "Known Recoverable Resources of Uranium 2007," www.world-nuclear.org/info/inf23.html.

Dangers

The first thing most people think of with regard to the dangers of nuclear power is the possibility of a rogue state or terrorist organization obtaining a functional nuclear bomb from a military stockpile. This is definitely a possibility, and a lot of effort is being devoted to ensuring that it does not happen. However, it has little bearing on commercial civilian nuclear power plants.

Still, there are several significant dangers associated with nuclear power. Radiation from spent fuel rods and waste material is a health hazard. Fuel rods and waste materials contain plutonium, which rogue states or terrorists might be able to use to make nuclear bombs if they got hold of it. Damage to nuclear reactors, which might be caused either by natural events such as earthquakes or by terrorist activity, is a concern because of the potential for releasing radioactive contaminants. Finally, damage to long-term storage facilities for radioactive waste material could cause severe health and environmental problems.

Nuclear power is widespread throughout the world in civilian power plants and moderately common on naval warships and on some civilian icebreakers. The nuclear industry has an enviable safety record. Except for the accident at Chernobyl, which I discuss shortly, no workers or member of the public have ever died from radiation from a commercial power plant accident.[23] When we look at industrial accidents from 1970 to 1992 in the United States and the United Kingdom and calculate the number of deaths per quantity of electricity produced, the number of deaths is small. For every death in nuclear power plants, there were 10 in natural gas plants, 43 in coal plants, and 110 in hydroelectric plants.

There have been only two serious nuclear power plant accidents worldwide: at Three Mile Island in the United States in 1979 and at Chernobyl in Ukraine in 1986. In the Three Mile Island incident, a cooling malfunction led to a partial core meltdown. According to the Nuclear Regulatory Commission, "The accident at the Three Mile Island Unit 2 (TMI-2) nuclear power plant near Middletown, Pa., on March 28, 1979, was the most serious in U.S. commercial nuclear power plant operating history, even though it led to *no deaths or injuries to plant workers or members of the nearby community.*"[24] There was a partial meltdown of the core and release of radiation. However, the core would have remained intact if not for operator error, and the radiation was contained by the reactor containment vessel. That is, even with the much-feared core meltdown, which was only partial in this case, the safety systems worked and prevented a serious situation.

The April 1986 accident at Chernobyl, Ukraine, was the worst nuclear accident anywhere. The explosion and resulting fire, products of a flawed Soviet reactor design coupled with serious mistakes made by the plant operators, destroyed the reactor. The major flaw was that there was no containment vessel surrounding the reactor, as has always been standard in US power plants and is now standard worldwide. According to the World Nuclear Association,

> The accident destroyed the Chernobyl 4 reactor, killing 30 operators and firemen within three months and several further deaths later. One person was killed immediately and a second died in hospital soon after as a result of injuries received. Another person is reported to have died at the time from a coronary thrombosis. Acute radiation syndrome (ARS) was originally diagnosed in 237 people on-site and involved with the clean-up and it was later confirmed in 134 cases. Of these, *28 people died as a result of ARS within a few weeks of the accident. Nineteen more subsequently died between 1987 and 2004 but their deaths cannot necessarily be attributed to radiation exposure. Nobody off-site suffered from acute radiation effects* although a large proportion of childhood thyroid cancers diagnosed since the accident is likely to be due to intake of radioactive iodine fallout.[25]

The death toll eventually rose to fifty-six, about half of the deaths occurring in the one or two days immediately after the accident. The health effect on the children is unfortunate. Radioactive iodine-131, one of the by-products of a nuclear reactor, is not especially dangerous unless ingested, in which case it accumulates in the thyroid and can lead to thyroid cancer. Thyroid cancer is easily treated if detected early and seldom leads to death. At Chernobyl, there were nine deaths from four thousand cases of thyroid cancer, a survival rate well above 99%. And since the half-life is of iodine-131 is only eight days, if consumption of milk from the affected area had been prevented for a month or so, the number of thyroid cancers resulting from the accident would have been greatly diminished.

The common concern that terrorists might attack a nuclear power plant is overstated. In the first place, nuclear power plants cannot explode like bombs. The physics simply does not allow that to happen. (This is not true of some advanced reactor design concepts on the drawing board, but it is definitely true of existing designs.) The main concern about terrorists attacking a nuclear power is that radiation could be released. However, power plant containment vessels are quite robust, and analyses show that they can withstand direct impact from a fully fueled 767 jet aircraft. Indeed, nothing short of repeated artillery assault or detonation of a bunker-busting bomb would cause release of any radiation. Overall, the release of significant amounts of radiation is extremely unlikely. Terrorists with a weapon that could breach a reactor containment vessel would be able to cause much more damage and loss of life using the weapon elsewhere.

Our experience with nuclear power since the dawn of the nuclear age and the lessons of Three Mile Island and Chernobyl show that, with appropriate attention to design, training, and operation, nuclear power plants are as safe as, or safer than, any other type of power plant.

Bottom Line

Nuclear power is a pollution-free source of electricity. Reserves should satisfy world demand for uranium for one hundred years. Beyond that, potential developments hold the promise for a virtually limitless supply of energy. Skyrocketing uranium prices in the past few years are reason for concern. However, the current price situation seems related to the cost of increasing production rather than to inadequate reserves. Increases in price should shortly stimulate increased production, which will eventually bring prices down.

US reserves of uranium are modest, and the United States has to import almost all of the uranium it consumes. Fortunately, the major sources are good friends. However, this might change in the future, and the small amount of domestic uranium will be a continuing concern. Developing other sources of uranium, such as fast breeder reactors and seawater, will alleviate the concern.

On the negative side, there is justifiable concern about safety of operation and safety of radioactive waste storage. However, the dangers are manageable. As real as they are, nuclear power health and safety hazards are generally overstated. Nonetheless, any effort to expand nuclear power must address the public's safety concerns.

Summary

The US energy situation is troubling, both from the point of view of domestic reserves and dependence on foreign sources. At the current rates of domestic production and consumption, US proved reserves of petroleum will last 7 years, of natural gas will last 12 years, and of coal and uranium will last roughly 225 years. Coal will last so long because the United States has huge reserves. Uranium will last so long because we import 80% of our consumption. Can we stop importing foreign oil, natural gas, and uranium? That is, can domestic reserves support current consumption for a reasonable time? At the present, the United States imports 66% of its oil, 16% of its natural gas, and 80% of its uranium. For domestic oil reserves to support consumption, domestic production would have to triple. If this were even possible, reserves would then last only three years. Production of natural gas would have to be increased 20%, which is reasonable, but domestic reserves would then last only ten years. Uranium production would have to be increased a staggering thirteenfold, resulting in a mere seventeen-year reserve.

The reserves will almost definitely last longer with the development of UTRR, but such development may take decades. Increasing oil and uranium production as much as suggested would be difficult. It seems unreasonable to expect that the United States will be able to stop importing energy as long as we continue to consume energy at current rates. Fortunately, world reserves of oil, natural gas, and uranium are larger than domestic reserves. The world will be able to draw on proved reserves of oil and natural gas for about 50 years, uranium for about 80, and coal for over 140 years at current rates of consumption. One should expect that with increasing population and increasing industrialization of the third world, demand for energy resources will grow apace. If consumption increases 3% annually, as did US consumption of oil over the past hundred years, the lifetime of these reserves decreases to a few decades.

Oil is unique in that it is the exclusive current source of vehicle fuel. At the same time, oil provides many products such as lubricating oils, fuel oils, and petrochemical industry feedstocks that we depend on. Conserving gasoline and diesel fuel will ease demand for oil somewhat, but we will still need oil as long as we have no alternative source for the other petroleum products.

Natural gas is almost as good a fuel as oil. Natural gas supplies heating, it is a potential vehicle fuel, and it is cleaner than oil. However, natural gas is also running out, and we will soon be in a situation similar to that in which we are with respect to oil now, so natural gas is not the final solution, just a resource that prolongs availability of petroleum.

Coal is a different story. US coal reserves will last 225 years at current rates of production and consumption, much longer than the 164 years other world coal reserves will last. Indeed, the US does not import any coal and holds a quarter of the world's reserves. When it comes to coal supplies, the United States is in a strong position. Unfortunately, coal power is polluting, and unless we learn how to clean up its production, depending on coal could be a health and environmental disaster. Moreover, coal is not a vehicle fuel, and unless we learn how to extract gasoline and diesel from coal, it will not help us with dwindling supplies of oil and natural gas. Alternatively, the United States could shift from an oil economy to an electricity economy, drawing on coal as a primary resource. The inherent inefficiency of power station turbines makes coal power's high pollution levels even more of a problem for electricity. We must reduce emissions of greenhouse gases and other pollutants and reduce damage to the environment. Otherwise, a major increase in coal consumption would not be acceptable.

Nuclear power promises a clean, virtually inexhaustible source of electricity. Uranium is abundant, and nuclear power would meet our growing demand for electricity as long as we manage the health and safety issues satisfactorily. As with coal, a shift from oil- to electric-powered vehicles could greatly increase demand for nuclear power. If we overcome safety concerns, nuclear power may be a virtually inexhaustible pollution-free source of electricity.

Much of the rest of this book deals with stretching oil reserves by making gasoline vehicles more efficient and replacing gasoline with some nonpetroleum fuel. But even if we learn to live without gasoline, we will not survive in a post-petroleum world unless we find other sources or replacements for other petrochemical products.

CHAPTER 2

Conventional Vehicles

Gasoline and diesel together are responsible for 23% of our energy consumption, 23% of our greenhouse gas emissions, and 64% of our petroleum consumption. Road vehicles consume almost all of this. Improving the fuel economy of highway vehicles is the most important single step in reducing our consumption of gasoline and combating global warming. We can do this by nontechnological means, such as driving less or adopting efficient driving behavior, or by technological means, such as making automobiles more fuel-efficient or replacing gasoline with some other fuel. This chapter examines current automotive technology and explores which fuel-economy improvements are practical and how much improvement is possible without radical shifts in automotive technology.

Road transportation started expanding rapidly around 1950 with the convergence of two trends: oil and the automobile. Development of the modern automobile dates to the late 1700s with a model-sized self-propelled vehicle, followed by pedal power and some modern features in 1780, the internal combustion engine (fueled by hydrogen and oxygen) in 1806, and an electric car in 1881. The modern automobile dates from 1885 when German inventor Karl Benz patented the four-stroke internal combustion gasoline-fueled automobile. Benz began to sell his vehicles in 1888. German engineer Rudolf Diesel patented the diesel engine in 1892 and built the first one in 1897. By 1910, the field had shaken out, and the four-stroke gasoline-fueled internal combustion engine had gained dominance over competing steam, electric, two-stroke gasoline, and diesel engines. The large-scale production line manufacturing of affordable automobiles was introduced by Ransom Olds in 1902 and expanded by Henry Ford in 1914. With the wide popularity of the internal combustion engine automobile, gasoline and diesel became the major products of crude oil.[1]

The availability of affordable automobiles and inexpensive gasoline greatly influenced development of roads, housing, and cities. Started in 1956, the interstate highway system enabled Americans to travel great distances inexpensively and conveniently and has guided construction of cities for the past fifty years. Inexpensive personal transportation has molded the American lifestyle, where we live, where we work, where we shop, how far we are willing to travel to visit relatives, and how far we are willing to travel for vacations. It is foolhardy to think that we are going to restructure all of this in a few years.

The Department of Transportation estimates that there were 254 million highway vehicles registered in the United States in 2007.[2] The Environmental Protection Agency (EPA) estimates that between 11 million and 13 million cars go to the scrap heap each year. That is, drivers take 5% of vehicles off the road every year, junk them, and replace them with new ones. The EPA estimates that 38% of the vehicles on the road are more than ten years old. If this replacement rate remains constant, even if an ultraefficient car becomes available tomorrow and only the ultraefficient cars are sold from then on, it would still be at least twenty-one years before the overall fuel economy of automobiles on the highway settled down to the new ultraefficient value. Significant change in gasoline consumption will take decades.

Although some vehicles are gas hogs, some very fuel-efficient cars are available now. Over the entire fleet of new cars, fuel economy ranges from 10 miles per gallon (mpg) to 46 mpg, with an average about 22 mpg. The average fuel economy has remained nearly constant for the past twenty years.[3]

In order to meet the long-range goal of becoming independent of foreign oil, we have to eliminate or substantially reduce demand for gasoline. Simply improving the fuel economy of standard gasoline vehicles will not achieve this goal. It will reduce demand for gasoline and postpone the inevitable depletion of oil reserves, but it will not eliminate demand. Only replacing gasoline with something else will do that. However, developing alternatives to the internal combustion gasoline engine will be a lengthy process. In the short term, all we can do is improve current vehicle technology so that we can stretch the limited reserves of gasoline as far as possible while we develop alternatives. In this chapter, I describe the sources of inefficiency and possible engineering means for improving the fuel economy of the internal combustion gasoline engine. I do not discuss diesel engines here, not because they are not relatively common, but because the difference between diesel and gasoline engines is not great enough to warrant special treatment at the level of the discussion in this chapter. I discuss diesel engines and other alternatives to current gasoline vehicles in the next chapter.

Automobile Basics

When sitting at a stop sign or traffic light or stuck in a traffic jam, you are not going anywhere. If the engine is running and consuming fuel, all of the energy in the burned fuel is lost. As you drive away from the stop sign, your engine burns fuel to accelerate to the desired cruising speed. Some of the fuel energy goes to overcome engine and transmission-system losses. The rest is expended in maintaining the moving vehicle's momentum. Heavier cars require more energy to accelerate than smaller cars and hence are less efficient. At cruising speed, the engine must provide enough energy to overcome external forces such as drag and tire rolling resistance in addition to overcoming the system losses to maintain the vehicle's momentum. When you stop, you must counter the vehicle's momentum by braking. Simply stated, braking increases frictional losses (brake shoes rub on brake drums; brake pads rub on brake discs), converting stored energy into heat, which air flowing over the braking system removes from the vehicle. All the energy that was stored in momentum is lost to heat the atmosphere.

The two diagrams in figure 2.1 show the losses in a typical internal combustion gasoline engine automobile. One diagram applies to city driving and the other to highway driving. Highway driving usually involves higher speed, less frequent acceleration and braking, and less idling. I have calculated engine loss assuming an air standard Otto cycle engine with 10:1 compression ratio and 1.4 specific heat ratio.[4] Other thermodynamic engine cycles are possible, but the Otto cycle is a good model for the typical automobile of the past decade. That is, the diagram is a good baseline for understanding recent engineering and estimating the potential for further improvement.

The single largest source of energy loss in a vehicle is the engine, which accounts for about 80% of the total loss. One might conclude that huge improvements in fuel economy are possible through engineering research and development (R&D). However, this conclusion is not valid. The thermodynamic efficiency of the Otto cycle internal combustion gasoline engine operating with typical air/fuel mixture and compression ratio is roughly 60%.[5] Reducing engine losses below 40% is simply not possible. Nonetheless, there is plenty of room for improvement. Most of the reducible losses are those resulting from less-than-ideal fuel delivery, fuel burning, and exhaust. The EPA estimates that engineering improvements to fuel delivery and burning can reduce engine losses by around 30%, although such reductions require increased complexity and result in greater cost. Good engineering, good manufacturing, and improved design features, such as variable valve timing and lift, cylinder deactivation, turbochargers and superchargers, and direct fuel injection all contribute to improved fuel economy. The rest of the losses are due to friction. While friction may be reduced through good engineering, it is always a factor and increases with engine size, because with larger engines come larger rubbing surfaces, and

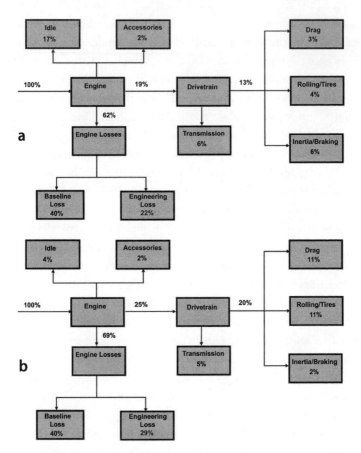

FIGURE 2.1   Energy Flow in Typical Midsize Gasoline Car: a, city driving; b, highway driving. These charts show where energy in the fuel is lost. The major loss is in the internal combustion engine and drivetrain. Some of this loss can be overcome by engineering improvements to the engine and the transmission. Some of this loss is inherent in the thermodynamics of the engine and cannot be reduced. Fuel waste during idling is a major source of loss in city driving, which could be overcome by turning the engine off when not needed. Major sources of loss in highway driving are aerodynamic drag, which could be addressed by streamlining, and tire rolling friction, which could be addressed by low-rolling-friction tires or simply by driving slower.

Sources: Tires and Passenger Vehicle Fuel Economy, Transportation Research Board Special Report 286 (Washington, DC: National Research Council, 2006), 40; Richard Stone and Jeffrey K. Ball, *Automotive Engineering Fundamentals* (Warrendale, PA: SAE International, 2004).

with speed, because driving faster means higher engine speed and thus higher engine friction. Engine friction can be reduced by making engines smaller and less powerful and by driving slower.

Idling losses are simply due to burning fuel to keep the engine running. After basic engine losses, idling is the leading loss factor in city driving, accounting for 17% of the overall loss. Turning the engine off when stopped would achieve huge savings, increasing typical fuel economy in city driving 4 mpg. Balancing the savings is the added fuel needed to restart the engine, the loss of engine-driven accessories while idling, and the inconvenience and delay of restarting. An integrated starter/generator system automatically turns the engine off at idle and starts it again with only a slight delay when the accelerator is pressed. Such a system could cut the losses at idle in half. Idling is not as large a source of waste in highway driving because there is less stopping. Nonetheless, eliminating idling would provide a noticeable improvement in efficiency even in highway driving.

Accessories such as lights, windshield wipers, and instruments are unavoidable. One can also argue that air conditioning is necessary. Heating is not really an accessory, as most cars use heat from the engine normally dumped into the atmosphere and lost. Radios and other electronics also consume energy. We cannot do much about the energy lost to accessories.

The main loss element in the drivetrain is the transmission. An internal combustion engine operates efficiently over a rather broad range of crankshaft rotation speed, roughly 600 to 7,000 revolutions per minute (rpm). The wheels on my car rotate between 0 rpm at idle and 1,000 rpm at 80 miles per hour (mph). The transmission is required to match the optimal engine speed to the vehicle's road speed. Transmission losses on the highway are generally slightly less than in city driving because most automobile transmissions provide optimum matching at the common highway cruising speed. For most cars built since the Arab oil embargo, the design speed is 55 mph. Manual transmissions may be up to 94% efficient, automatic transmissions as low as 70%. Of course, the efficiency of a manual transmission in practice depends on the driver's gear-shifting skill. The EPA estimates that using continuously variable transmissions or automatic manual transmissions, essentially a standard transmission shifted by the machinery rather than by the driver, could improve fuel economy 6%.[6]

The energy needed to overcome inertia and accelerate to cruising speed increases with heavier cars and higher cruising speed. It is more significant for large vehicles on the highway because weight and speed are higher. Making cars lighter and reducing cruising speed mitigate some of the negative effect of inertia.

Large frontal area, boxy design, and high speed increase aerodynamic drag and the associated energy loss. Streamlining the car body, making the car small, and driving slower reduce drag and improve fuel economy. The speed penalty is severe. While drag increases as the square of velocity, the power required to overcome drag increases as the third power of velocity. That is, doubling speed requires an eightfold increase in power. Driving 70 mph requires twice the power and twice the fuel as driving 55 mph. A small streamlined car generally has much less drag than a large boxy one, but limiting speed has much more effect and is easier to achieve.

Tires get hot when we drive because of resistance as the tires roll over the road. Low-rolling-resistance tires minimize rolling resistance and losses but give a harsher, less comfortable ride.

Where does this leave us with regard to improving fuel economy? Many aspects of the engine are amenable to engineering advances, and this is where most of the effort is going. Engine designers have a good grasp on friction, and I would not expect much improvement here. All we can do is reduce friction by reducing the size and speed of the engine. Losses at idle can be reduced by turning off the engine when the car is not moving, despite some inconvenience. High-tech transmissions, such as the continuously variable transmission or automatic manual transmission, can decrease transmission losses markedly, although they do cost more. Inertia is not amenable to change in the internal combustion engine (though regenerative braking in electric-drive cars can recoup something). Reducing frontal area, streamlining the car, and reducing top speed all reduce drag losses. High-tech tires can reduce, though not eliminate, rolling resistance at the expense of a harsher ride.

How much improvement can we expect? First, let us look at engineering tweaks. Assume that good engineering eliminates three-quarters of the engine engineering losses, reducing engineering losses in city driving from 22% (as shown in fig. 2.1) to 5% and total engine loss from 62% to 45%. This is about a 30% reduction in engine loss as estimated by the EPA. It also means that the engine would be operating close to the theoretical maximum possible efficiency, which is a stretch, and leaves little room for further improvement. Then assume that three-quarters of transmission loss is eliminated by high-tech transmissions, three-quarters of tire loss is eliminated using low-rolling-resistance tires, and three-quarters of the drag loss is removed with streamlining. Overall, this would eliminate 26% of the losses in city driving and 39% of the losses in highway driving. That is, fuel economy of a typical 20 mpg car would increase to 27 mpg city and 33 mpg highway, close to the fuel economy of the Chevrolet Aveo: 27/34/31. (In this standard way of representing fuel economy, the first number is mpg in city driving, the second is mpg in highway driving, and the third is a composite number assuming 55% city driving and 45% highway.)[7] We have already made most of the possible engineering improvements to the internal combustion gasoline engine and automobile. There is not much room for additional improvement as long as we continue with the standard internal combustion gasoline engine.

CAFE Standards

Congress established the Corporate Average Fuel Economy (CAFE) standard in 1975, in response to the 1973 Arab oil embargo.[8] CAFE is the average fuel economy of a manufacturer's fleet of passenger cars and light trucks for the current model year. The goal of the original CAFE standard was to increase new-car fuel economy to 27.5 mpg by model year 1985. The CAFE standard has changed several times since it was established. The current standard is 27.5 mpg for cars and 22.2 mpg for trucks with a gross vehicle weight rating (GVWR) of 8,500 pounds or less. Vehicles with GVWR greater than 8,500 pounds (i.e., the large sport-utility vehicles [SUVs]) were exempt.

In May 2009, President Obama raised the fuel economy standard. Under the new rules, the mandated average economy of each manufacturer's fleet of new cars and light trucks increases to 35.5 mpg (39 mpg for cars, 30 mpg for trucks), with the increase being phased in between the 2009 and 2016. This is very laudable and a huge step in the right direction, but we have to understand the details.

First, the improvement will not happen overnight. It will be seven years before the standard reaches 35.5 mpg, and it will take at least two decades from then for standard-satisfying fuel-efficient vehicles to replace all cars on the road. Figure 2.2 illustrates the difference between the CAFE standard that applies to new vehicles and the average fuel economy of actual vehicles on the road. The figure shows the CAFE standard and actual fuel economy over the past thirty years. Although the CAFE standard has been 27.5 mpg since 1985, actual fuel economy has only reached 22.4 mpg. The figure also shows my projection of how the actual fuel economy will respond to the 2009 CAFE standards. I assume that we continue to replace 5% of vehicles on the road each year with vehicles meeting the then-current fuel economy. The year 2006 was the last year for which actual data were available, so the projection starts in 2007. The projection shows that it will be 2053, almost forty-five years from now, before we achieve 34 mpg actual fuel economy, still short of the 35.5 mpg CAFE standard.

Second, the CAFE standard will not accomplish its goal if the public does not buy the cars. One estimate says that the new efficiency standards will add $1,300 to the price of each car, on av-

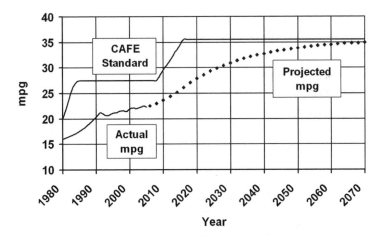

FIGURE 2.2 Average US Fuel Economy. This chart shows how slowly actual fuel economy of cars on the road responded to CAFE standards over the past thirty years since CAFE standards were instituted. The chart also shows the projected fuel economy response to new standards, assuming a 5% replacement rate. Although the CAFE standard rises to 35.5 mpg in 2016, on-road fuel economy will not reach 34 mpg until at least 2053. That is, oil consumption would be reduced about a third in fifty years if the total number of cars on the road does not increase and the annual replacement rate remains at 5%.

Source: Bureau of Transportation Statistics, *National Transportation Statistics*, table 4-23, "Average Fuel Efficiency of US Passenger Cars and Light Trucks," www.bts.gov/publications/national_transportation_statistics/html/table_04_23.html.

erage. Fuel cost savings will recover some of this, but the amount of saving will depend on the price of gasoline. If the benefit is too little, people may not continue to replace cars at the 5% rate. Consequently, the projection may be optimistic, and it might be more than thirty years before the on-road fleet is completely converted to fuel-efficient vehicles.

Third, the CAFE standard will not accomplish its goal if manufacturers do not comply. Even though manufacturers pay stiff penalties for not meeting the standards, several manufacturers do not meet the current CAFE standards. They have decided that it is better for them economically to pay the penalty than to invest in improving fuel economy. European manufacturers consistently pay millions of dollars in penalties a year. Asian and most large domestic manufacturers usually pay no penalty.

Finally, raising the actual fuel economy from 22.4 to 34 mpg reduces gasoline consumption by only 34%, barely enough to cut our foreign oil imports in half.

The new CAFE standard is a step in the right direction and is achievable, but we need to do much more.

EPA Fuel Economy Ratings

This is a good time to describe how EPA determines fuel economy. The test vehicle is put on a machine called a dynamometer that simulates the driving environment, just as exercise bikes and treadmills simulate physical activity. The dynamometer controls the resistance provided by the rollers under the drive wheels, simulating acceleration, hills and so on, while the operator controls the speed to follow the established test protocol.[9] The amount of fuel consumed during the test is the fuel economy for that particular vehicle model and test schedule. The EPA used only city and highway test schedules up to 2007. Three additional schedules (high speed, air conditioning, and cold temperature) are used today in an attempt to get fuel economy ratings that are better matches to actual highway performance. Figure 2.3 shows the highway test protocol.

The purpose of the EPA fuel economy ratings is to determine whether manufacturers satisfy the CAFE standards. The ratings do not necessarily say much about the gasoline mileage a driver should get from a car, and the ratings are inadequate for comparing different engine technologies. Let me explain.

The dynamometer test is a strictly defined and controlled test procedure. Because the EPA rates all cars with the same test schedule, one can compare test results and say that a certain model is better than another model, slightly better, not as good, and so on. But the test cannot account for how a particular person drives (a

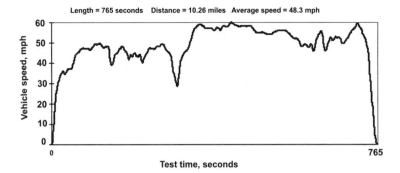

Length = 765 seconds   Distance = 10.26 miles   Average speed = 48.3 mph

Test time, seconds

FIGURE 2.3   EPA Highway Fuel Economy Test Protocol. The vehicle operator maintains the vehicle speed specified in the test protocol while the dynamometer simulates road incline by providing the desired resistance.

Source: EPA, "Detailed Test Information," www.fueleconomy.gov/feg/fe_test_schedules .shtml.

heavy foot on the accelerator, poor gear shifting skill, etc.), how much excess cargo they have in the car adding to weight, how the car is maintained, what the driving environment is (temperature, snow, rain, etc.), and so on. The EPA rating for a gasoline car is only a rough guideline for actual fuel economy. However, the actual fuel economy of two models with the same EPA rating should be roughly the same as long as the same driver operates them with the same level of maintenance, loading, environment, and so forth. Recently defined additional test protocols attempt to match common driving conditions to give a better indication of actual fuel economy and quiet consumer complaints about EPA fuel economy ratings.

We must be careful about comparing different automobile technologies using the fuel economy ratings. The following three examples should be thought-provoking. The Volkswagen Jetta, discussed in more detail in the next chapter, is a good example. The EPA rating for the diesel version is 36 mpg, while the rating for the gasoline version is 26 mpg. Concentrating on the EPA ratings, one might get the impression that the diesel is better than the gasoline model. However, when we consider consumption of crude oil, the source of both gasoline and diesel, the result is different. Although the diesel consumes less refined fuel than the gasoline model, a barrel of crude oil has less diesel fuel than gasoline, and the diesel consumes more crude oil than the gasoline vehicle. If the chief concern is consumption of crude oil, the gasoline Jetta is more crude-efficient than the diesel model. I return to this topic in the next chapter when I discuss diesels.

Flex-fuel vehicles burning E85 ethanol fuel (consisting of 85% ethanol and 15% gasoline) provide a second example. According to the EPA ratings, flex-fuel vehicles generally get 25% poorer fuel economy than straight gasoline vehicles. In particular, a midsize Dodge Avenger gets 23 mpg on gasoline and 17 mpg on E85. However, if we are concerned about consumption of gasoline and crude oil, one should consider the composition of E85 fuel. Since only 15% of it is gasoline, 17 miles per gallon of E85 is equivalent to 113 miles per gallon of gasoline. While the EPA has a procedure for adjusting fuel consumption ratings for flex-fuel vehicles, the purpose is to document adherence to the CAFE standards and does not necessarily indicate how much gasoline one is using or saving.

My last example is the electric vehicle. General Motors has stated that EPA tests show that the Chevrolet Volt gets 230 mpg,[10] which is a truly impressive fuel rating. However, the EPA test protocols are not set up to handle cars that get some motive power from batteries and some from gasoline, as is the case with the Volt. A more realistic figure is 50 mpg. I discuss the Volt's fuel economy in detail in chapter 3.

Nonengineering Approaches to Fuel Economy

What can we do to save fuel without buying high-tech cars? First, we can drive fewer miles. But driving less might be difficult because decades of cheap fuel have led to the development of our highway system and separation of living and working centers, requiring long commutes, having to drive children to activities, and so on. Moreover, driving less consumes less gasoline but does not improve fuel economy per se. Second, we can drive smaller, lighter, less powerful cars. There is a clear advantage to doing so, but Americans must overcome their love of macho driving. Third, we can drive smarter. The EPA website lists several techniques for driving efficiently:

- Stop aggressive driving. This could improve fuel economy 5% to 33%.

- Drive at good speed. Fuel economy suffers significantly at speeds slower than 25 mph or faster than 55 mph. Save 7% to 23%.

- Reduce what you carry in your car. Each one hundred pounds of load changes fuel economy by 2%.

- Reduce excessive idling. Seventeen percent of fuel energy is lost to idling in city driving.

- Use cruise control and overdrive gear on the highway.

- Keep tires properly inflated. Fuel economy suffers 0.3% for each 1 psi difference from optimum in all four tires.

- Keep the engine tuned up. Not doing so can cost 4% in fuel economy.[11]

These EPA guidelines are understandable in light of figure 2.1, showing energy losses in the internal combustion engine. Aggressive driving puts a lot of energy into vehicle momentum that is lost if it has to be removed by braking. Acceleration is essentially inefficient. High speed contributes directly to air resistance and indirectly to engine friction losses because high road speed requires high engine speed. The effect of weight is obvious, as greater load requires more work. Keeping tires inflated properly minimizes the frictional losses in tire rolling resistance. Idling is also an obvious loss.

There seems to be no consensus about how much the average driver can improve fuel economy by practicing all of the EPA suggestions, especially because many of us already follow some of the recommendations. Still, widespread good driving habits could markedly raise national average fuel economy. A recent fad indicates what is possible. "Hypermiling" is the name given to driving techniques that markedly increase fuel economy.[12] Some of these techniques, such as turning the engine off and coasting down inclines are dangerous and not to be encouraged. Still, hypermilers have claimed fuel economies of 76 to 213 mpg. The success of the fad shows strongly how much driving-behavior modification can influence fuel economy, much more in some respects than engineering developments can.

Summary

Examining energy losses for a typical late twentieth-century gasoline automobile provides insight into where R&D efforts should go to improve efficiency. While there is room for improvement, and raising the CAFE fuel economy standard is definitely a step in the right direction, it will take over forty-five years for the average fuel economy of cars on the road to rise to 34 mpg, 1.5 mpg less than the new CAFE standard. Indeed, if the new fuel-efficient cars cost more than current vehicles, there will be a resistance to purchasing new cars, and it will take longer for the fleet of cars on the road to transition to higher economy. The efficiency of current conventional gasoline automobiles is probably about as good as we can expect from internal combustion engines. There is not much more room for engineering improvement, and additional improvement will depend on further reducing size, weight, power, and drag. Overall, we have gotten about as much out of the internal combustion gasoline engine as possible. Meeting the new CAFE standard will require some additional engineering improvement in gasoline automobiles, balancing less fuel-efficient vehicles with smaller and lighter cars and adding advanced technology to the mix. Further improvement will require different technology. Even if we meet the 35.5 mpg goal, it will be about 2070 before we do so. Moreover, increasing from 20 mpg to 35.5 mpg only reduces demand for gasoline 43%, not nearly enough by itself to eliminate the need to import foreign oil. We need additional improvements, such as the alternative vehicles discussed in the next chapter.

Distinct from increasing fuel economy by engineering changes, we can decrease demand for gasoline by modifying our behavior. Driving less and driving less aggressively would each have a marked effect on gasoline consumption. However, I am not sanguine about our doing so. The American lifestyle has developed over the past hundred years based on ready availability of inexpensive gasoline and the acceptance of a lot of driving. Changing this will not be easy. Many Americans are habituated to aggressive driving and will not give up such habits without a struggle. It may be possible to mandate sedate driving by enforcing limits on vehicle size, speed, power and so on, but any effort to do so would encounter substantial resistance. Requiring a 35.5 mpg CAFE standard for some vehicles while exempting gas-guzzlers also will not accomplish our goals.

Much of the improvement in fuel economy comes from reducing automobile size and weight, and many people are rightly concerned about the safety implications of mixing small fragile cars with large, heavy SUVs, trucks, and vans. Numerous studies, including recent crash tests of the very small Smart car, verify that the small cars do not do well in crashes with large cars. It is essential that efforts to make our highways safer for small cars accompany any efforts to downsize automobiles.

CHAPTER 3

Green Vehicles

The previous chapter dealt with reducing gasoline consumption without major technology changes. We can drive less; we can drive more efficiently; we can drive smaller, lighter automobiles. We can also get some improvement in fuel economy by making mechanical improvements to the internal combustion gasoline automobile. Doing these things will decrease gasoline consumption somewhat but not as much as desired. They will not achieve the goal of eliminating oil as a fuel source and replacing it with cleaner domestic and possibly renewable sources. To accomplish this, we need some serious engineering changes.

This chapter focuses on technology solutions for increasing gasoline fuel economy and for replacing gasoline with alternative fuels, either completely or partially. Each alternative has potential benefits and certain drawbacks and limitations. Each alternative breaks new engineering ground and has a large price tag. Understanding the engineering possibilities and trade-offs is important if we are to make the tough decisions we must face in the future. I discuss engineering alternatives that are available now or are under development: diesel, flex-fuel, natural gas, hybrid electric, series electric, plug-in electric, and hydrogen fuel-cell vehicles.

The first three types of alternative vehicles I discuss, diesel vehicles, flex-fuel vehicles, and natural gas vehicles (NGVs), are similar in that they have internal combustion engines for propulsion. The differences among them are the fuel and the fact that the "fuel tank" on the NGV is one or more compressed gas cylinders. There are several electric vehicles (EVs), which get some or all of their propulsion from an electric motor. The hybrid electric vehicle (HEV) is unique in that it has two parallel propulsion systems, one an internal combustion engine and the other an electric motor. The series electric vehicle (SEV) has all-electric drive but burns fossil fuel in an onboard generator; the generator is in series with the electric motor. The hybrid plug-in electric vehicle (H-PEV) is similar to the SEV, but its batteries can be charged from the power grid. These three are transitional vehicles between fossil-fuel internal combustion engine vehicles (ICEVs) and pure electric drive. The final two vehicles, the pure plug-in electric vehicle (P-PEV) and hydrogen fuel-cell vehicle (HFCV), are pure electric-motor propulsion systems that get all of their energy from the power grid.

Diesel Vehicles

Diesel engines are similar to gasoline engines in that they are powered by internal combustion and run on petroleum-based fuel. They are different in that compression rather than an electrical spark ignites the fuel and the fuel comes from a different fraction of crude oil. Diesel vehicles have been on the road for almost as long as gasoline engines, but they have generally been restricted to trucking in the United States because of a perception that they are noisy and polluting. However, diesels tend to be dependable, long lasting, powerful, and fuel-efficient, and they have been much more popular for personal automobiles in Europe for some time. Diesels have become reliable, efficient, quiet, and low-pollution, and they have been steadily gaining acceptance for automobiles in the United States. Diesel vehicles are roughly 30% to 35% more fuel-efficient than gasoline engines, and diesel fuel contains 10% more energy per gallon than gasoline. Modern diesels are attractive because fuel economy is better than with gasoline. Even though diesel fuel costs more than gasoline, the better fuel economy results in lower fuel cost per mile.

The functional layout of the diesel vehicle is shown in figure 3.1a. Similar to the conventional car discussed in chapter 2, the diesel vehicle has a fuel tank filled from an external source, an internal combustion engine, and a drivetrain that incorporates a transmission. Several models of diesel vehicles are currently available in the United States. The Volkswagen Jetta SportWagen is a useful example because it is available in otherwise identical diesel and gasoline versions. The diesel SportWagen has the leading fuel economy in the small-station-wagon category: 30 mpg in city driving, 41 mpg in highway driving, and 35 mpg overall (assuming 55% city driving and 45% highway driving); written succinctly as 30/41/35. This compares favorably with the gasoline version at 21/30/25. The diesel gets 40% better overall fuel economy.

The diesel Jetta produces lower greenhouse gas emissions than the gasoline version. I calculate 38% less CO_2 from the diesel than from the gasoline Jetta (3.0 tons per year versus 4.8 tons per year). Clearly, the Jetta diesel reduces CO_2 emissions per mile significantly compared with the gasoline version.

The saving in annual fuel cost is attractive. While diesel fuel costs more than gasoline, the Environmental Protection Agency (EPA) estimates the diesel Jetta saves $210 a year in fuel costs. However, the purchase price of the car offsets these benefits. The sticker price of the Jetta diesel is $4,600 higher than the gasoline model. Higher sticker price is common for alternative vehicles. In the case of the Jetta, it would take twenty-two years of fuel-cost saving to recoup the increased vehicle cost. On a simple financial-investment basis, the diesel does not make sense at current prices. It might be desirable for reducing pollution and fuel consumption, but there is little financial incentive for the average person to invest in the Jetta diesel.

When we look beyond fuel economy, the overriding issue is reducing demand for crude oil and dependence on imported oil. This leads to an interesting situation. A barrel of crude oil contains less diesel fuel than gasoline, about half as much. When we calculate the fuel economy as *miles per barrel of crude*, rather than *miles per gallon of refined diesel fuel*, the result is 338 miles per barrel of crude for the diesel versus 500 miles per barrel of crude for the gasoline version. That is, the diesel consumes more crude oil per mile than the gasoline version, even though it consumes less refined diesel fuel. As long as the number of diesels on the road remains small, modifying the mix of crude oils and adjusting the refining process can offset at least some of the difference. Modifying international trade in refinery products would also help. However, as the number of diesels on the road increases, the demand for crude oil will increase, just the opposite of what we are trying to accomplish.

Biodiesel

Biodiesel provides a possible way out of this unacceptable situation. Biodiesel fuel can run a diesel engine, but it comes from biomass rather than petroleum. From here on, I distinguish between "biodiesel," which comes from plants and animal fat, and "petrodiesel," which is the common petroleum-based diesel fuel.

Biodiesel is one of two biological alternatives to fossil fuel. I discuss ethanol, the leading contender for a possible alternative to gasoline, in the section on flex-fuel vehicles. In both cases, diluting petrofuel with biofuel reduces oil consumption. Whether this is successful depends on how the alternative fuel performs in the vehicle and the inevitable drawbacks introduced by the biofuel.

Ironically, when Rudolf Diesel was first working on his diesel engine invention in the late nineteenth century, he envisioned using vegetable oil as a fuel so that fuel for his engine would be available throughout the world. One demonstration engine at the 1900 Paris Exposition ran on peanut oil. The burgeoning petroleum economy and easy availability of kerosene and, later, diesel fuel pushed vegetable oil fuels into the background.[1] Even as late as 1912, shortly before his death, Diesel was extolling the virtues of vegetable oil as a fuel for his engines. Diesel's vision not withstanding, kerosene from refining petroleum was plentiful and inexpensive in the late nineteenth century, and Diesel concentrated on kerosene as

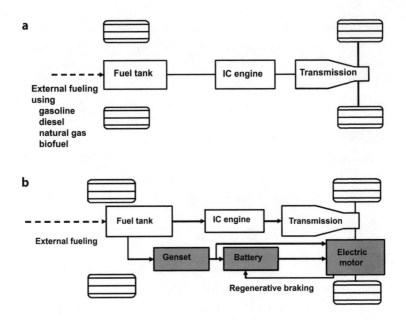

FIGURE 3.1   Fossil-Fuel Vehicle Layouts: a, conventional internal combustion engine vehicle (ICEV); b, hybrid electric vehicle (HEV); c, series electric vehicle (SEV); d, hybrid plug-in electric vehicle (H-PEV). The ICEV layout illustrates the common automobile with a fuel tank, internal combustion (IC) engine, and transmission. Liquid fossil fuel, gasoline or diesel, is stored in a simple fuel tank in conventional ICEVs. NGVs are ICEVs that use compressed natural gas stored in high-pressure (3,600 psi) cylinders for fuel. The HEV has two complete propulsion systems. One is the common internal combustion engine; the other is an electric drive system. The small battery is charged by regenerative braking or a small on-board genset. The SEV has a single propulsion motor, which is electric. Its small battery is charged by regenerative braking and an onboard genset. It does not have the fossil fuel IC engine of the HEV, and the battery cannot be charged by plugging into an external charging source. The H-PEV is similar to the P-PEV (see fig. 3.3), which gets all of its electricity from an external charging source, but unlike the P-PEV, the H-PEV has an onboard genset to provide electricity after the battery has depleted the external charge.

Source: Mehrdad Ehsani, Yimin Gao, and Ali Emadi, *Modern Electric, Hybrid Electric, and Fuel Cell Vehicles: Fundamentals, Theory, and Design*, 2nd ed. (Boca Raton, FL: CRC Press 2010).

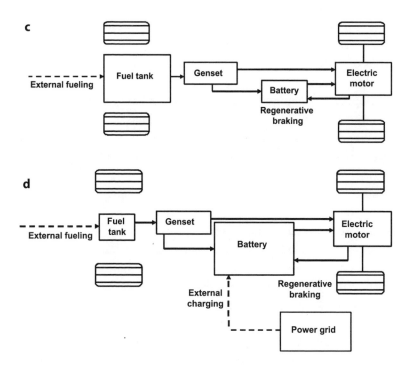

c

External fueling → Fuel tank → Genset → Battery → Electric motor

Regenerative braking

d

External fueling → Fuel tank → Genset → Battery → Electric motor

External charging

Regenerative braking

Power grid

a fuel. By the early twentieth century, diesel engines had become more widely used, and an inexpensive, low-grade petroleum product called "diesel fuel" had become available. Diesel engines were optimized to take advantage of this new fuel, making it more difficult for them to burn vegetable oil. Even then, people were concerned because petrodiesel fuel was dirtier than vegetable oil fuels, but petroleum was inexpensive and plentiful, and petrodiesel became the standard fuel.

Vegetable-oil fuel was not totally forgotten. It is clean, and vegetable oil sources are widely available in tropical climates. Research continued in the early twentieth century. The advent of World War II and the resulting disruption of transoceanic transportation and petroleum availability stimulated a resurgence of interest in vegetable oil as a biodiesel fuel, but interest again flagged after the war when petroleum supplies recovered. Biodiesel experienced another temporary resurgence because of the 1973 OPEC oil embargo. Another period of resurgence is now underway because of the pending crises of petroleum supply and global warming.

The current situation is slightly different from when Diesel was developing his engine. His original engine could burn vegetable oil directly. Since then, however, petrodiesel has become available, and engineers have optimized the diesel engine to take full advantage of the new fuel. One result was that vegetable oil, or any other biofuel, does not work as well in diesels now as it did then. The solution is either to modify the engine to burn biodiesel more efficiently or modify the biodiesel fuel. Industry's choice has been to modify the fuel. The reason for this decision is telling. Essentially the industry concluded that modifying machine tooling and engine design and manufacturing would not be economically beneficial because biodiesel would never capture a significant fraction of the diesel market. The cost-effective approach is to modify the fuel to match the engine rather than the engine to match the fuel, with the result that producing biodiesel fuel is more complicated now than it was in Diesel's time. The diesel industry itself does not think that biodiesel can be a major contributor to the diesel engine industry.

The industry is currently extracting biodiesel from plants, used vegetable cooking oil, and animal fats. I will not elaborate on the process because there are several books that cover the topic thoroughly,[2] and I am much more interested in production yields. With regard to plant sources, the plant has to be grown and harvested, and the important statistic is yield of biodiesel per acre of agricultural land (table 3.1).

The most common biodiesel crops in the United States are soybeans and corn, which provide very little biodiesel. Other crops grown in other countries with different climates and agricultural conditions might be more productive, but relying on them would raise the same concerns about dependence on foreign sources of fuel as we now have about foreign oil. Rising demand for biofuel has led to an increased amount of arable land being devoted to producing fuel rather than food. This means less food and higher food

TABLE 3.1   *Biodiesel Crop Yields*

Crop	Yield (gallons per acre)
Oil palm	635
Coconut	287
Rapeseed, canola	127–160
Peanut	113
Sunflower	102
Soybean	48
Corn	18

Source: Greg Pahl, *Biodiesel: Growing a New Energy Economy*, 2nd ed. (White River Junction, VT: Chelsea Green, 2008), chapter 3.

prices worldwide and has led to widespread deforestation throughout the world, with a detrimental effect on global warming.

The potential beneficial effect of biodiesel production from US agriculture is inevitably limited. As stated earlier, the United States consumes 64.3 billion gallons of diesel and 142 billion gallons of gasoline (208.3 billion gallons of fuel total) annually. If 10% of the 922.1 million acres of US farmland were devoted to soybeans, the yield of biodiesel would be 4.4 billion gallons, only 7% of the annual diesel consumption, or 2% of total vehicle fuel. Actual savings would be slightly different because biodiesel would be blended with petrodiesel in vehicle fuel, but the numbers are telling. The point is that taking a large percentage of US farmland out of growing food would make only a minor contribution to oil independence.

Biodiesel can also be produced from used cooking oil. The quality is not as good as biodiesel derived from plant crops because of contaminants and the successive rounds of heating to high temperature. However, at 50% yield, the 3 billion gallons of used cooking oil generated in the United States annually could theoretically provide 1.5 billion gallons of biodiesel. That is only 2% of diesel fuel demand. We could also get 750 million gallons of biodiesel (1% of US consumption of diesel) from 11 billion pounds of animal fat available from the meat-packing industry. All told, converting all the used cooking oil, all the animal fat, and 10% of

US farmland to biodiesel production would barely offset 10% of current petrodiesel consumption. Moreover, increased demand for diesel would simply mean that biodiesel would provide a smaller percentage of total consumption of diesel.

I am uneasy depending on crops for fuel. What happens if we depend on the US soybean crop for fuel and the country enters a period like the 1930s dust bowl? Crop yields already vary from year to year, and now, with increasing concern about global warming and climate change, we should expect crops to be less dependable. To be somewhat whimsical, what happens if Americans stop eating french fries and the supply of cooking oil dries up? What happens if the United States turns vegetarian and the supply of animal fat dries up?

I cannot leave biodiesel without mentioning algae. Algae potentially surpass all other biodiesel feedstocks with regard to yield per acre. Early experiments by the National Renewable Energy Laboratory (NREL) indicate that maintaining high levels of production is difficult, but yields of 6,500 gallons per acre are possible.[3] This is several hundred times the yield from growing soybeans. Experimentation ended because of high cost and difficulty getting consistent dependable yields. However, interest is increasing again, and algal biodiesel warrants careful monitoring. Inexpensive dependable algal biodiesel with yields anywhere close to the promise could be the breakthrough that weans us from gasoline.

Bottom Line

Diesel vehicles reduce greenhouse gases and pollution and increase fuel economy. However, they also increase demand for oil. Should the number of diesels on the road grow significantly, we will have to increase imports of crude oil.

In the short term, increased sticker price more than offsets the saving in fuel expense. This will probably change in the future, but now buying a diesel car is not a wise economic decision.

Domestic agricultural biodiesel cannot provide enough fuel to be a significant player in the long-term national energy crisis. Biodiesel has many advantages, and developing a dependable supply of biodiesel would certainly help to stretch fuel supplies, but the overall effect would be minor unless algal biodiesel becomes a reality. Algal biodiesel holds promise, and we should watch its development closely. The game changes significantly if a dependable, economical, source of algal biodiesel with yields anywhere near what seems to be feasible becomes available. In that case, diesel engines running on biodiesel would be a major benefit. Even then, however, depending on a crop could be risky. We should scrutinize dependability of crop yield very carefully.

Flex-Fuel Vehicles

During World War II, the fuel industry added tetraethyl lead to gasoline to boost octane. Lead turned out to be toxic and was banned. Several other additives for boosting performance were tried at various times. One of these, methyl tertiary butyl ether (MTBE), was added to gasoline to boost octane and oxygenate the fuel to make it burn more efficiently. This was successful, and the Clean Air Act of 1990 established MTBE as the standard additive. Unfortunately, MTBE turned out to be a carcinogen, and it was replaced by ethanol, an alcohol derived from corn. Since 2007, federal law has required all "gasoline" to be 10% ethanol and 90% gasoline, the mixture referred to as E10. While originally a performance-boosting additive, ethanol is now being considered for diluting gasoline. The idea is to replace 85% of the gasoline in each gallon of fuel with ethanol by mandating E85 fuel. However, ethanol is highly corrosive and, although E10 is not a problem for engines, E85 would damage conventional gasoline engines. Engines have to be modified to burn E85. A flex-fuel vehicle is capable of burning any mixture of gasoline and ethanol, at least up to E85. The concept is alluring: replace most of the gasoline consumed in vehicles, much of which comes from foreign oil, with domestic, renewable, corn alcohol. Unfortunately, the flex-fuel vehicle is promising in principle but problematic in execution.

One problem has to do with compression ratio. Ethanol has higher octane than gasoline. Indeed, it is used as an additive to gasoline to raise the octane. As long as the flex-fuel engine has to accommodate gasoline, the compression ratio has to be low, in keeping with gasoline. If engines burned only E85, the compression ratio could be higher, taking advantage of the higher octane of ethanol, and efficiency would be better. That is, keeping the engine flexible enough to burn either gasoline or E85 reduces the efficiency of burning E85. That said, there are several other reasons ethanol fuel is not a good idea.

Consider the effect on food supply. If we assume a corn-ethanol crop yield of 700 gallons per acre, devoting 10% of the 922.1 million acres of US cropland to ethanol production would provide 64.9 billion gallons of ethanol. Because driving range with E85 is 66% of the driving range with gasoline, this amount of ethanol would offset 30% of the gasoline consumed in the United States annually. That is, devoting a tenth of existing cropland to ethanol production would offset 30% of our gasoline consumption, whereas the same acreage devoted to biodiesel would offset only 7% of diesel consumption. Clearly, ethanol reduces gasoline demand more than biodiesel does, and devoting cropland to ethanol production would have a greater effect on reducing oil consumption.

Ethanol production that does not affect the food supply may be possible. Cellulosic ethanol is produced from wood, grasses, or the nonedible parts of plants. Ethanol may also be obtained from algae. Algenol Biofuels, Inc., is working closely with Mexican industry and the Mexican government to develop the technology.[4] Construction of the pilot plant began in January 2010.[5] The potential yield of ethanol from algae is similar to the potential yield of biodiesel from algae, up to 6,000 gallons per acre, so these sources might eventually produce ethanol economically and without negatively affecting the food supply

However, in addition to having the same negative effects on world food supplies and deforestation as biodiesel production, reliance on ethanol has several unique and severe drawbacks. First, ethanol contains less energy per gallon than gasoline (table 3.2). The range of a flex-fuel vehicle operating on E85 is three-quarters the range when operating on gasoline. This is an inconvenience to drivers on long trips, and experience in other countries where ethanol fuel is more widely used shows that many drivers switch back to gasoline on long trips to minimize the frequency of stops for refueling. Availability is another issue. As of late 2010, there were 2,347 filling stations in the United States that sold E85.[6] There are no significant technological hurdles to overcome in producing E85, but it will be many years before E85 is widely available. Cost is another issue. E85 costs more per gallon than gasoline.[7] Yielding fewer miles per gallon at a higher price per gallon, E85 costs almost twice as much per mile traveled than does gasoline. Operating a flex-fuel vehicle on E85, one would have to fill up more often and pay more per mile—not an attractive proposition. While cost will probably decrease with time and research, it is doubtful that range, which is limited by tank size and the lower energy density of ethanol, will ever be as great as we are used to with gasoline.

Another problem with ethanol is that it is highly corrosive. The current automobile engine and fuel system cannot stand up to high concentrations of ethanol. E10 is not harmful to current automobiles (boats and small airplane are another matter, discussed below), but E85 definitely is. Any metal, rubber, or fiberglass component that is subject to ethanol's corrosiveness must be replaced to run on E85.[8] Flex-fuel vehicles are engineered to burn E85 or any combination of E85 and gasoline without damage. This means that if E85 is mandated, not only all automobiles but all gasoline engines, including those for motorcycles, small boats, small airplanes, all-terrain vehicles, standby home generators, snowblowers, lawnmowers, power tools, and so on will have to be replaced with E85-tolerant models to avoid damage. The cost to the public

TABLE 3.2   *Energy Densities of Fuels*

Fuel	Gravimetric		Volumetric		gge (gallons)
	MJ/kg	kWh/pound	MJ/L	kWh/gallon	
Gasoline	46.4	5.8	34.2	36.0	1.0
Diesel	46.2	5.8	37.3	39.2	0.92
Biodiesel	42.2	5.3	33.0	34.7	1.04
Ethanol	30.0	3.8	24.0	25.2	1.43
E10	43.5	5.5	33.2	34.9	1.03
E85	33.1	4.2	25.6	26.9	1.34
Natural gas	53.6	6.8	0.0364	0.04	939.56
Hydrogen	143.0	18.0	0.0108	0.01	3,166.67

Source: Wikipedia, "Energy Density: Energy Densities Table," http://en.wiki
pedia.org/wiki/Energy_density.
Note: A gge (gallon of gasoline equivalent) is the amount of fuel needed to
provide the same energy as one gallon of gasoline.

would be enormous. If we were to mandate the use of E85 in automobiles and keep E10 available for other uses, ensuring that E85 is
always used in automobiles would be difficult. Drivers would tend
to use E10 because of its lower cost and greater driving range.

The boating industry provides indications of what might happen if there were a widespread switch to E85. Powerboats have had
problems ever since 2007 when E10 was mandated. The first problem is that the corrosive ethanol fuel dislodges gunk from the walls
of the fuel tank and fuel lines and dissolves fiberglass fuel tanks.
The resulting contaminated fuel clogs fuel filters and causes engine
stoppage and sometimes engine damage. The second problem is
caused by ethanol's affinity for water. Water in the fuel can cause
a phase separation, with gasoline on top and a mixture of ethanol
and water at the bottom. The ethanol/water blend may cause the
engine to stop and may cause severe damage to the engine.[9] Boat
fuel tanks are usually made of fiberglass and are vented to the atmosphere, and they go long periods without use, during which the
environment provides many opportunities for water to enter the
fuel tank. The problems caused by E10 in gasoline engines other
than those in modern automobiles are so severe that Oregon passed

a law that went into effect January 1, 2009, exempting gasoline sold for use in boats, aircraft, all-terrain vehicles, classic cars, and gasoline-powered tools from the requirement to contain 10% ethanol.[10] E10 is a problem for boating and small aircraft now. E85 will be a much more serious problem for more engines in the future.

While it is true that flex-fuel vehicles burning E85 would reduce gasoline consumption and demand for foreign oil, the negative effects are numerous and must be weighed very carefully. First, the driving range of E85 is significantly less than that of gasoline, and cost per mile is much higher. Getting people to switch from gasoline to E85 will be difficult. Second, extracting ethanol from corn, the current plan, would have a significant effect on world food supplies. In the broad scheme of things, people will accept ethanol only if we find a source that does not affect food supplies or lead to deforestation. Yields from cellulosic sources are probably too low to be useful, but this might change. Ethanol from algae might be the solution, but that is a long way off. Even if we find a source of abundant ethanol that does not affect the world food supply, there is the third problem: what to do with nonautomobile engines, such as small boats, small airplanes, snowblowers, and power tools. Turning them into flex-fuel engines would be expensive, and maintaining distinct supplies of E85 and gasoline (or E10) would be a logistical nightmare.

Natural Gas Vehicles

Although only one natural gas car is on the market as of this writing, other NGVs, such as buses and delivery trucks, are moderately commonplace. An internal combustion engine can burn natural gas as a fuel with relatively minor modifications. Fuel economy and cost per mile are similar to conventional vehicles. Moreover, natural gas burns cleaner than gasoline. NGVs definitely reduce emissions of greenhouse gases and pollution and certainly reduce demand for gasoline. In-vehicle fuel storage and refueling are the primary drawbacks.

The NGV layout is shown in figure 3.1a. The NGV is an ICEV similar to the common gasoline car, but the fuel is natural gas compressed to 3,600 psi, and the fuel is stored in a high-pressure cylinder.

A little arithmetic clearly illustrates the storage issue. The energy content of natural gas is 0.04 kWh per gallon at atmospheric pressure (table 3.2). Compared with gasoline's 36.0 kWh per gallon, this is minute. One would need 940 gallons of natural gas at atmospheric pressure to provide the energy in one gallon of gasoline. To be practical in a vehicle, natural gas has to be stored in a much smaller volume.

There are two ways natural gas can be stored efficiently. The first method is cooling and liquefying it as liquefied natural gas. This provides good energy density by volume but requires special tanks to maintain very low temperature. This is expensive and therefore practical only for long-distance transportation, such as shipping, and for heavy-duty vehicles, such as large trucks, but it is impractical for private passenger vehicles. The second method of storing natural gas efficiently is to compress it. This is straightforward and requires nothing more complicated for storage than a strong storage tank and a compressor for filling the tank. When natural gas is compressed and stored at 3,600 psi, the standard for natural gas, the energy content is still only about a quarter of the energy in an equivalent volume of gasoline. In other words, to store the same amount of energy in natural gas as in gasoline, one needs about four times as much storage volume to get the same driving range as with gasoline. Moreover, the high pressure requires a strong, heavy tank. Natural gas storage takes up much more space and weighs much more than gasoline storage for equivalent quantities of energy.

How do natural gas cars stack up to gasoline cars? The Honda Civic is a good illustration because it comes in a both a gasoline model (the DX) and a natural gas model (the GX). Figure 3.2 com-

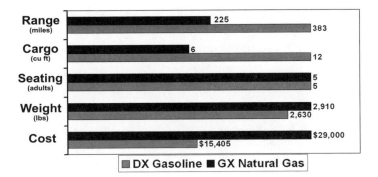

FIGURE 3.2 Honda Civic Gasoline DX versus Honda Natural Gas GX. Although the gasoline and natural gas Honda Civic cars are otherwise similar, the natural gas model costs much more, weighs more, has half the cargo space, and has about half the range.

Source: Honda, "Civic," http://automobiles.honda.com/civic/.

pares the two. Predictably, the natural gas version weighs 10% more than the gasoline model, its range is 40% less, and its cargo space is 50% less. There also is a price penalty. While fuel economy is slightly better, 31 mpg of gasoline equivalent (mpgge) versus 29 mpg, and fuel savings are about $100 per year, the sticker price is $14,000 higher. It would take 140 years to recover the purchase price penalty. As we have found with all of the alternative technology vehicles, there is no financial incentive for selecting an NGV over the gasoline car. The main benefits are that natural gas is much less polluting than gasoline, emitting roughly 20% less pollution overall and 15% less CO_2, and switching to NGV reduces consumption of gasoline and oil.

Sizing the fuel tank is a compromise between driving range and cargo space. To provide the same range as the gasoline model, the natural gas model would have to have a gas tank four times the size of the gasoline tank, undoubtedly taking away useful cargo space. Honda has decided to provide about half the range of the gasoline model, a barely acceptable 225 miles, with a gas tank about twice the size of the gasoline tank. Range is lost and some cargo space is lost.

Fueling an NGV requires filling the high-pressure natural gas tank, and how this is done depends on what is available and the complexity of the fueling station. Home refueling is possible with a "Phill" system.[11] The Phill system allows one to refuel at home from the household gas line. One must have a natural gas line and a garage to house the system, so availability of home refueling is very limited. Most townhouse and apartment dwellers would be out of luck. Another problem is that residential gas lines provide gas at low pressure, and this means the system has to compress the gas. Compressing heats gas, so the gas must not only be compressed but cooled. This is not difficult, but it takes time. Anyone familiar with filling SCUBA tanks or barbecue grill gas tanks knows the drill. Specifications for the Phill system indicate a fueling rate of about half a gallon of gasoline equivalent per hour, which is one hour for 12 miles of driving and nineteen hours to fill the Honda GX tanks. The cost of purchase, installation, and electric power to run the system are prohibitive. Home fueling is available now for the few people who have natural gas service and a garage, do not mind paying dearly for fuel, and do not object to overnight fueling. Widespread home refueling of private NGVs is prohibitively impractical and expensive, although it does not require infrastructure expansion.

At fueling stations equipped with high-pressure storage tanks, filling the car's tank is fast and simple. The gas is precompressed and stored at high pressure. Filling the tank is simply a matter of connecting the high-pressure storage tank to the vehicle tank and letting the gas flow. Unfortunately, there are very few such "fast-fill" fueling stations. This will change in the future if demand for NGVs grows. However, given the lower range of NGVs, the network of fast-fill stations will have to be more extensive than the current network of gasoline stations. Building the necessary infrastructure will be slow and expensive.

The need for heavy high-pressure fuel tanks on the vehicle is a limiting factor for small cars. On smaller vehicles, the percentage of weight and volume devoted to the fuel tank increases. Consequently, natural gas is most appropriate for large vehicles. Application to fleet vehicles like buses and delivery vans is ideal. Because they are large vehicles, the fuel tank weight and size penalty is proportionally less. Because they could operate out of a central depot, they would need only a single fueling station, minimizing the required infrastructure development.

If we are to introduce NGV nationally, we need a national infrastructure for fueling. Since the range of passenger NGVs is about half that of gasoline vehicles (based on the Honda GX NGV as a model for future NGV cars), the number of natural gas refueling stations probably needs to be at least twice the number of gasoline stations. Considering that there are few if any public natural gas refueling stations outside of California (as of 2010), this is a tall order.

Should NGVs become commonplace throughout the country, demand for natural gas will increase, and we will have to work toward doubling the supply of natural gas and distribution pipelines. The effect on natural gas imports will be significant. We consume about 23.2 Tcf of natural gas annually. Most but not all is from domestic sources. As we saw in chapter 1, we import 16% of the natural gas we consume in the United States. The natural gas equivalent to all the gasoline consumed in the United States is 15.8 Tcf. This means that demand for natural gas would increase almost 70% if NGVs became the standard mode of transportation. If domestic production remains constant, the additional 15.8 Tcf would have to be imported, raising the amount of natural gas we import to 50% of consumption. This is a long way from the goal of minimizing dependence on foreign sources of energy.

Refueling an NGV at home is difficult, time consuming, expensive, and available only to people who have parking immediately adjacent to their homes and a natural gas line. Home refueling of NGVs would be impractical for most people and too expensive for all. Widespread NGV deployment will have to wait for the infrastructure of fast fueling stations to be developed, and the cost will depend on the size and complexity of the new infrastructure. Since the range of current NGVs is significantly shorter than we are used to with gasoline cars, there would have to be significantly more fast fueling natural gas stations than the current number of gasoline stations if we convert passenger cars from gasoline to NGVs. Infrastructure development would be complex and costly. If, on the other hand, we use natural gas only for fleets of vehicles that return periodically to a central depot for fueling, far fewer natural gas fueling stations would be needed. The complexity and cost of infrastructure development would be less and the development time shorter.

A massive shift from gasoline to natural gas would reduce greenhouse gases and pollution and would decrease demand for gasoline and oil, but it would increase our dependence on foreign natural gas and hasten the time when we start experiencing shortages. Limiting natural gas to fleet vehicles would prolong our domestic reserves and minimize demand for imported natural gas and might achieve a better balance of gasoline and natural gas usage.

Finally, the size and weight penalty imposed by natural gas storage tanks limits our ability to improve fuel efficiency by making vehicles small and light. There is some vehicle size where the penalty associated with natural gas offsets the potential benefits of size and weight. The relative weight/size penalty is less prohibitive for fleet vehicles than for small cars.

For these reasons, the future of NGVs appears to lie in large vehicles, such as large taxis, buses, delivery vans, and other large fleet vehicles, fueled at central depots. The tank size/weight penalty is not severe for these larger vehicles, and refueling at a central fleet depot minimizes how much the natural gas infrastructure has to be expanded.

Hybrid Electric Vehicles

One has only to mention the Toyota Prius to get an earful about how good gasoline HEVs are in terms of fuel economy. Indeed, in every category of passenger car or light truck tabulated by the EPA in its *Fuel Economy Guide 2009* that includes an HEV, the HEV is the fuel-economy leader. The 2009 Prius, which has a 48/45/46 mpg rating, has led the hybrids several years running.

Why is the HEV so successful, and what improvements may we expect from HEV technology in the future? What does HEV technology tell us about the goals of driving green in general?

"Hybrid" means that the vehicle uses more than one source of energy. The primary source for HEVs is gasoline, and the secondary source is electricity. That is, a standard gasoline engine is the primary source of propulsion, and an electric motor powered by a battery augments the gasoline engine in certain situations. Figure 3.1b shows the layout of an HEV, emphasizing the two parallel propulsion systems that make the vehicle a hybrid. The battery for the electric drive is charged either by regenerative braking or by the motor-generator set, or genset, which burns whatever fuel is used for the internal combustion engine propulsion system.

The HEV provides benefit in four ways. First, hybrid vehicles utilize varying numbers of elements of the improved standard technology already incorporated in conventional cars, such as a continuously variable transmission, cylinder control to improve efficiency of the engine and drivetrain, low-rolling-resistance tires, and streamlining to reduce losses. Though not strictly hybrid technologies, as they are common in standard gasoline engines as well, these elements improve fuel economy in both city and highway driving. Second, the electric motor comes on at high speed under certain circumstances to assist the gasoline engine when high power is required. This improves fuel economy slightly during highway driving because it allows the gasoline engine to be smaller and more efficient than otherwise would be possible (the 2010 Toyota Prius gasoline engine is a mere 98 horsepower). Third, the HEV almost eliminates idling loss by turning off the gasoline engine when the vehicle is not moving and turning it back on when the vehicle starts moving again. The integrated starter/generator technology makes this possible with no hesitation because the system uses the electric motor and battery for getaway while the gasoline engine starts and comes on line. Fourth, HEV technology reduces inertia losses by using regenerative braking to capture the inertia that otherwise would dissipate to the atmosphere as heat. The reclaimed inertia energy charges the battery and makes the whole electric motor scheme possible in the absence of external charging. Regenerative braking and the integrated starter/generator are the major unique contributors to HEV fuel economy, and both are possible only because of the electric propulsion system. Advanced internal combustion engine engineering is applicable to conventional automobiles and is not unique to HEV; streamlining and weight reduction are also applicable to conventional automobiles, and they benefit small cars operating at slower speeds more than they benefit large, boxy trucks and sport-utility vehicles (SUVs).

In 2007, I contemplated trading my Toyota Highlander in for a new model. Since an HEV model was available at the time, I seriously considered getting one. I didn't. My experience provides a good example because the 2007 Toyota Highlander was available in otherwise similar HEV and conventional models. EPA fuel economy figures are 32/27/29 for the Highlander hybrid and 19/25/21 for the conventional version. Note that hybrid technology provides a lot of benefit in city driving (32 mpg versus 19 mpg) but not very much on the highway (27 mpg versus 25 mpg). Hybrid technology benefits highway driving mainly by using the electric motor to assist the gasoline engine at high speed and during acceleration. However, the atmospheric drag induced by the large, boxy SUV body overpowers the benefit of the hybrid assist in highway driving. The noticeable benefit in city driving is most certainly due to regenerative braking and eliminating idle losses. HEV technology would have saved me about $524 a year in fuel cost.[12] However, the hybrid version cost $10,000 more than the gasoline version. It would have taken twenty years to recover the purchase price penalty. I eventually decided to keep what I had. The improvement in fuel economy achieved by hybrid technology on a large, boxy vehicle was not worth the added cost.

Results from similar comparisons of smaller cars are different. The EPA *Fuel Economy Guide 2009* shows, in particular, the top conventional and top hybrid cars in the compact category.[13] Here the conventional Chevrolet Aveo gets 25/34/29 mpg, and the Honda Civic HEV gets 40/45/42 mpg. The benefit of the HEV over a conventional competitor is 45%, which is double the 21% improvement of the 2007 Highlander HEV over the conventional model. However, comparing the overall fuel economy of the conventional Chevrolet Aveo with the conventional Jeep Compass, one sees a 19% improvement, about the same as the improvement from the conventional 2007 Highlander to the hybrid 2007 Highlander. If I want to improve fuel economy, then, I can do as well or better by switching to a smaller conventional car than by switching to a hybrid in the same category, and it will cost me a lot less. Once again we see that the losses incurred by large, heavy, boxy vehicles are simply too severe to be overcome by technology.

Many of the engineering modifications incorporated in HEVs address engine and transmission inefficiency, air resistance, and tire rolling resistance common to all ICEVs. What distinguishes the HEV is a second, electric propulsion system in addition to the gasoline engine. The electric motor provides propulsion assistance at high speed, enables regenerative braking, and eliminates losses when idling. Most of the advanced engineering is just as beneficial on conventional vehicles, so it is not entirely clear how much the HEV-specific elements improve fuel economy relative to the added complexity and expense.

At the end of chapter 2, I estimated the likely maximum fuel economy for a standard gasoline engine automobile by looking at the energy flow (fig. 2.1) and estimating potential improvement in each area of loss. To do the same with an HEV, I assume that in addition to all of the improvements of the conventional vehicle, the integrated starter/generator saves 90% of the idling loss, regenerative braking recovers 85% of the inertial loss, and streamlining recovers 50% of the drag losses. The resulting estimate is 48 mpg in city driving and 43 mpg in highway driving. This is almost exactly the fuel economy for the 2009 Prius (48 mpg city and 45 mpg highway). My conclusion is that hybrid technology, like the latest standard engine technology, has already achieved about as much improvement as one can expect.

Comparing HEVs with standard cars shows the true cost. I compared vehicles in three categories. In the compact category, I compared the 2009 Honda Civic standard car with the Civic HEV. In the midsize category, I compared the latest Toyota Prius with the Nissan Versa. In the SUV category, I compared standard and HEV versions of the Ford Escape. In all categories, the HEV got better fuel economy and saved between $430 and $707 per year in fuel cost. Countering this saving was a price difference ranging from $8,000 to almost $16,000. It would take eighteen to twenty-two years before the increased purchase price is balanced by the fuel saving.

The 2010 Toyota Prius represents about as much as we can expect from engineering development of HEVs. Further improvement in fuel economy will consist of incremental improvements, such as using solar panels to help power accessories (as on the 2010 Prius) and making the vehicle lighter. However, it seems likely that the two-propulsion-system design will be superseded by mechanically simpler systems.

Electric Vehicles

The HEV is a very complex machine, having two parallel propulsion systems. Sometimes the gasoline engine propels the vehicle; sometimes the electric motor propels it; sometimes both propel it simultaneously. Most of the propulsion comes from the gasoline engine, with assistance from the electric motor. The HEV has allowed us to develop electric drive technology and demonstrate its advantages. The next conceptual development stage is the EV. These have been around for years for low-speed, short-distance applications like golf carts and factory runabouts. The design of current EVs suitable for highway use draw from experience with golf carts (and similar vehicles) and the HEV. The advantages of EVs are numerous. Electric motors are more efficient than internal combustion engines, they emit no greenhouse gases or pollution (though generating the electricity that drives them is another matter), they eliminate losses at idle, they minimize braking loss with regenerative braking, and they generally do not burn fossil fuel. Though some EVs do burn fossil fuel, P-PEVs do not.

Compare the diagram of the P-PEV in figure 3.3a with the HEV in figure 3.1b. Electric propulsion, battery charging from the power grid, and extreme simplicity are the key features. A P-PEV is plugged into an electrical outlet to charge the battery and runs on battery power until the battery is completely discharged. Table 3.3 summarizes the battery issues with two leading plug-in electric vehicles (PEVs), the Tesla Roadster and the Chevrolet Volt.[14] (The Chevrolet Volt is not a P-PEV, as I will discuss later.) The battery on the Chevrolet Volt weighs 375 pounds and has a 16 kWh storage capacity. It is capable of storing enough energy for 40 miles of driving before it has to be recharged. The Tesla Roadster gets greater range, 240 miles versus 40, with a larger battery. The larger battery on the Tesla Roadster does not completely account for the much greater range. To prolong the life of the battery, the Chevrolet Volt does not allow its battery to discharge completely and does not charge it fully. Only 8.8 kWh, roughly half of the total capacity, is used. The Tesla Roadster, on the other hand, uses the full capacity. The Tesla Roadster gets much greater range than the Chevrolet Volt but uses a much larger and heavier battery and costs much more.

There are a few engineering hurdles to overcome. The first is range, which is a critical issue. Chevrolet has decided that 40 miles per charge will satisfy most people. The argument seems to be that the average person drives 15,000 miles per year, which is just slightly greater than 40 miles per day. However, studies have shown that most Americans want 300 to 400 miles between fueling stops.

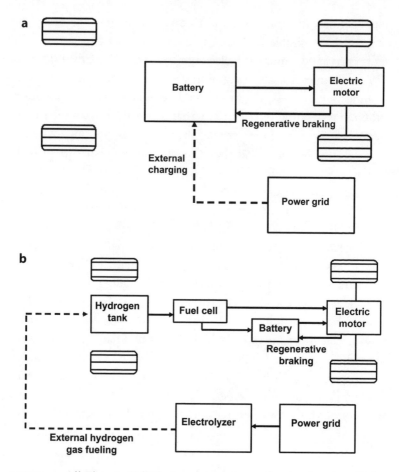

FIGURE 3.3 All-Electric Vehicle Layouts: a, pure plug-in electric vehicle (P-PEV); b, hydrogen fuel-cell vehicle (HFCV). The P-PEV gets all of its propulsion from an electric motor and all of its electricity from an externally charged battery. The HFCV also gets all of its propulsion from an electric motor but almost all of its electricity from the fuel cell, supplemented by regenerative braking. Consequently, the battery can be smaller than on a P-PEV. HFCVs would be fueled most practically at fast fueling stations that would generate hydrogen on site by electrolysis, drawing on power from the electric power grid and storing high-pressure hydrogen to facilitate fast fueling.

Source: Ehsani, Yimin, and Emadi, *Modern Electric, Hybrid Electric, and Fuel Cell Vehicles.*

TABLE 3.3   *Electric Vehicle Range Factors*

Item	Chevrolet Volt	Tesla Roadster
Expected price	$40,000	$109,000
Battery weight	375 pounds	990 pounds
Battery capacity	16 kWh	53 kWh
	(8.8 kWh usable)	(all usable)
Driving range	40 miles	240 miles
Battery economy	4.6 miles/kWh	4.5 miles/kWh

Sources: Chevrolet, "2011 Volt," www.chevrolet.com/volt/; Tesla Roadster, www.teslamotors.com/.

The Tesla is much closer to this goal than the Volt. Greater range means more battery capacity, which implies more batteries with greater volume and greater weight. There is a trade-off. Greater weight in batteries means less efficient driving and greater range. There is a break-even point somewhere. Research and development (R&D) may improve matters, but batteries are not new technology, and achieving the eightfold improvement required to raise range to 320 miles with little or no price increase is unlikely.

Charging the battery is the second engineering issue. Charging time is limited by the current capacity of the charging circuit and by the maximum charging current the battery can accept without damage. Much has been said about the simplicity of charging the battery by plugging it into a household electric circuit, but the physics of electricity dictate that a common household circuit can only provide the Volt 40 miles of driving from seven hours of charging; a kitchen small-appliance circuit improves this to five hours (table 3.4). A high-power electric stove or central air-conditioner circuit improves charging time to slightly more than one hour, but very few people have a spare high-power circuit, and installing one is expensive. Home recharging will be limited to drivers with off-street parking who can live with recharging times measured in hours. Most apartment dwellers and townhouse residents will be out of luck. Recharging P-PEVs will become practical only when stations capable of high-power fast charging are at least as widely available

TABLE 3.4   *Electric Vehicle Battery Charging Time*

Circuit	Current (amps)	Voltage (volts)	Power (kW)	Time to charge for 40 miles of driving
Standard household	15	110	1.65	7 hours
Household small appliance	20	110	2.2	5 hours
Central AC; electric cook stove	50	220	11.0	1 hour
Hypothetical fast charge station	—	—	3,000	13 seconds
Fueling from a 6-gallon-per-minute gasoline pump, assuming 30 mpg	—	—	—	13 seconds

Note: Charging times are calculated from the specifications for the Chevrolet Volt, which draws 8.8 kWh from the battery to drive 40 miles. Because of charging system losses, 11 kWh of electrical energy has to be put into the battery to provide 8.8 kWh. The hypothetical fast charge station is not practical because battery chemistry limits the maximum charging current and charging a fully discharged battery takes between one and three hours.

as the 159,006 gasoline filling stations in the United States.[15] The ideal fast charging station would provide electrical energy at the same rate that filling stations pump gasoline so that the time spent refueling an EV is the same as people are accustomed to. However, to charge the battery in the same amount of time that it takes to pump gasoline, one needs a 3,000 kW (3 MW) supply. Charging five EVs at the same time (most gasoline stations can fuel several cars at once) requires around 15 MW. This is a lot of power. To put it in perspective, the average capacity of US power plants is 62 MW;[16] the entire output from an average power plant would be needed to power four fast-charge EV charging stations. While the foregoing analysis is indicative of the huge power levels needed for fast charging, such extremely fast charging is impractical because the battery cannot accept such huge currents. Optimum battery charging applies the maximum charging current to the battery for about an hour and then a decreasing current for roughly two additional hours.[17] The limitation on charging current means that charging a fully discharged battery in less than one hour is not practical.

The third engineering hurdle is the effect on the national power grid. If we assume that fast charging stations are to be as widely available as gas stations, meeting the demand for electricity would be a big strain. First, consider the effect on the local distribution system. If I were to drive a Volt, I would consume 19 kWh of electricity per day. Reviewing my household bills, I currently consume 30 kWh per day on average. The electric car would increase my household consumption of electricity by 63%. A large number of people in my neighborhood suddenly doubling their consumption would severely strain the local distribution system. At the very least, the local distribution system would have to be upgraded as electric cars are deployed. Then consider the nationwide demand for electricity. Annual consumption of gasoline in the United States is equivalent to approximately 4.3 trillion kWh. Because of the efficiency of P-PEVs, they would require about half this, 2.42 trillion kWh of electricity. Now the annual US consumption of electricity is 3.5 trillion kWh. In order to replace all gasoline usage with electricity, we would have to almost double generating and distribution capacity. Doubling electricity consumption would also require a massive increase in the national power grid, especially transmission lines. There is already strong public resistance to new transmission lines; doubling the capacity would face strong opposition. More important is the need to double generating capacity. This would require massive construction of power-generating stations. Most important, most power plants now run on coal. Doubling the generating capacity with coal plants to support EVs might reduce dependence on oil, but it would also lead to unacceptable increases in pollution and damage to the environment. The P-PEV itself might be efficient and pollution-free, but generating the electricity to charge PEV batteries is less efficient and more polluting than gasoline.

The limited range of a P-PEV raises the question, what do you do if you run the battery dry? There is now no way to recharge the battery away from home and no means of getting road assistance.

Chevrolet's solution to this problem is to provide a small internal combustion engine motor and generator and a small fuel tank, variously referred to as a genset (motor-generator set)[18] or auxiliary power unit. Figure 3.1d shows the layout of the H-PEV, which is powered by the power grid or an onboard generator. The sizing is such that the generator keeps the battery charged after the initial 40-mile charge has been depleted, resulting in an overall 300-mile range. The internal combustion engine is small and can run at constant optimal speed, so the overall efficiency is quite good. Chevrolet advertises 50 mpg for the Volt in gasoline-generator mode.

There is a lot of talk about how much battery R&D will improve the practicality of electric cars. EVs probably will become more practical, but we need to go slow here. R&D may improve cost, weight, size, and longevity, but charging time depends on the power available at the plug. Available charging power is the limitation, not the battery.

Fuel Economy

Comparing fuel economies of EVs and gasoline vehicles is complicated because one uses electricity and the other burns gasoline. The common method for comparing fuel efficiencies using the gallon-of-gasoline-equivalent concept works well for fossil fuels but gets a bit more complicated when we compare fossil fuels and electricity. Energy content values are easily compared. A gallon of gasoline contains 36.0 kWh of energy, so we can treat every 36.0 kWh of electric energy consumed as the equivalent of 1 gallon of gasoline. That is, 36.0 kWh is one "gallon of gasoline equivalent," or 1 gge. An EV that drives 1 mile consuming 36.0 kWh of electric energy is getting 1 mpg of gasoline equivalent, or 1 mpgge. This works quite well if we think of electricity as a primary energy resource like wind or solar power. It gets complicated when we consider that electricity is actually a secondary energy resource generated from a fossil fuel in a power plant. One then has to take into consideration the conversion efficiency of the generator, be it the power plants feeding the national power grid or onboard gensets.

As I write this, I am watching a GM executive on television announce that the Chevrolet Volt will get 230 mpg fuel economy.[19] This is an incredible claim, and it makes the Volt the first car to claim triple-digit fuel economy. The claim is correct but misleading. Here is a good example of the need to understand the facts in order to understand claims like this about fuel efficiency.

To get this fuel economy rating, technicians subjected the car to the standard EPA test schedule of 11 miles and found that it consumed 0.22 gallons of gasoline. They then calculated fuel economy by adding the 11 miles of the test to the 40 miles the Volt would get from a fully charged battery, for a total of 51 miles, and dividing that by the 0.22 gallons of gasoline consumed by the genset. This is 230 mpg.

What is the real fuel economy of the Volt? It depends on how you determine fuel economy. One way is to consider only operation powered by the genset. Here, the small gasoline tank and genset power the car for about 260 miles. Chevrolet claims fuel economy of 50 mpg when operating on the genset. Another way is to consider pure battery operation. The fully charged battery carries the car 40 miles. The battery capacity is 16 kWh, but to extend the life of the battery, control circuitry never allows the battery to discharge or charge fully. The useful energy in a single charge is 8.8 kWh. Since 36.0 kWh is the equivalent of 1 gallon of gasoline, this is 0.24 gge, so fuel economy is 163 mpgge. If we want to account for the fuel required to generate the electricity needed to charge the battery, the PEV does not fare so well. Because of losses in the battery charger, it must draw 11 kWh from the grid to get 8.8 kWh into the battery. Since an average power plant is one-third efficient, we need to put 33 kWh of energy into the power plant to get 11 kWh into the grid. Overall, this is 0.92 gge, so battery operation gives 44 mpgge. This tells me that the PEV gets 44 to 50 mpgge regardless of fuel.

Additional improvements are possible with EVs. Figure 3.4 shows the energy flow in an EV and is similar to figure 2.1, showing energy flow in a gasoline vehicle. Losses are allocated to drag, rolling resistance, and inertia/braking according to the percentages from figure 2.1.

Compare the energy flow and losses in cars with electric drive shown here with that of conventional internal combustion engine cars shown in chapter 2 (fig. 2.1). Electric motors are more efficient than internal combustion engines, and the electric car does not require a complicated transmission. This results in drag and weight being more important relative to the other losses. What stands out is that much more of the energy in the battery gets to road contact (46% city, 67% highway) with EVs than with internal combustion engines. This means that reducing drag, tire rolling resistance, and inertia have much more effect on fuel economy in electric cars than in gasoline cars. Indeed, if one assumes that drag and inertia are reduced 75% by extreme streamlining and weight reduction and tire rolling resistance is reduced 50%, one would eliminate 30% of the losses in city driving and 45% of the losses in highway driving. Fuel efficiency would almost double on the highway and improve by 50% in city driving. Such measures will have a major effect on extending the range of PEVs.

Percentage losses are quite different once the internal combustion engine is replaced by the electric motor. Indeed, the losses attributable to drag, tires, and inertia/braking are much larger factors. This means that making vehicles smaller, lighter, and more streamlined should have proportionally greater effect on fuel efficiency in EVs than in gasoline vehicles.

Batteries

A key component of EVs (especially P-PEVs and H-PEVs) is the battery. This raises several concerns. The first is size and weight. Batteries are large and heavy relative to the energy they can store. This is not a major problem for HEVs, since large storage capacity

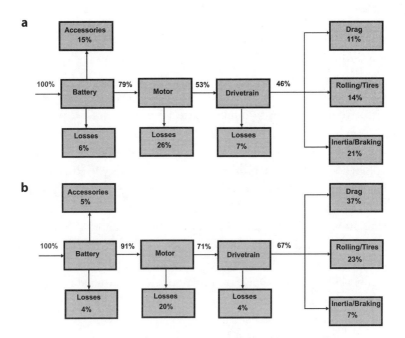

FIGURE 3.4 Energy Flow in an Electric Vehicle: a, city driving; b, highway driving. These models, illustrate energy flow in P-PEVs (i.e., vehicles with all-electric drive powered by a large battery charged from the power grid). When we compare these diagrams with similar diagrams for gasoline cars (fig. 2.1), we see that the electric vehicle has zero idle loss, extremely low engine losses, and no thermodynamic losses. Electric motors are more efficient than internal combustion engines, and the electric car does not require a complicated transmission. As a result, drag and weight are more important relative to other losses for electric vehicles. Much more of the energy in the battery gets to road contact with electric vehicles (46% city, 67% highway) than is the case with internal combustion engines. This means that reducing drag, tire rolling resistance, and inertia have much more effect on fuel economy in electric cars than in gasoline cars.

Source: Based on data from Kurt M. Johnson, "A Plug-In Hybrid Electric Vehicle Loss Model to Compare Well-to-Wheel Energy Use from Multiple Sources" (master's thesis, Virginia Polytechnic Institute and State University, 2008), 43, table 16.

is not required. It is a major problem for PEVs, since the range of the car depends on the size of the battery. Lithium-ion batteries are the most common EV battery technology today. The Chevrolet Volt battery pack weighs 375 pounds and takes up 100 liters (L), or 22 gallons, volume. The battery has a 16 kWh capacity, so the effective energy density numbers are 94 watt-hours (Wh) per kilogram and 160 Wh/L. Both are less than nominal values for the lithium-ion battery alone,[20] indicating that the battery system involves a lot of machinery in addition to the battery itself. The Volt has a range of 40 miles from a power source weighing 375 pounds and taking up a volume of 22 gallons. A standard gasoline car getting 30 mpg would need 1.3 gallons of fuel weighing 8 pounds to cover the same distance. The Volt suffers a 50:1 weight penalty and a 20:1 volume penalty compared to gasoline. Herein is one of the drawbacks of PEVs. One can only hope that continued R&D will get the size and weight down in order to provide adequate driving range in an acceptable package.

The second concern is air temperature. Prius owners in the Washington, DC, area have told me that there is a noticeable decrease in fuel economy in the winter. This is understandable because battery capacity depends on temperature and decreases if the temperature is lower than 75°F. Capacity decreases to 80% of maximum at 32°F and to 50% of maximum at 5°F. That simply means that the range of the Volt drops from 40 miles to 20 miles at temperatures approaching zero. To protect the battery, the battery pack in the Chevrolet Volt disconnects if the temperature gets down to around freezing. This could be a serious issue for people living in cold climates. They would need to keep the car plugged into the electric grid to keep battery temperature ready for instant departure. If a connection to the grid is not available, one would have to consume battery power continuously to keep the battery warm and ready for instant departure or wait while the battery warms up. In any case, maintaining battery temperature consumes energy, thereby reducing overall fuel economy. Operation in areas of the country subject to near-freezing temperatures reduces performance and subjects the driver to inconvenience.

The third concern is the supply of lithium, the fundamental raw material for EV batteries. There are other battery technologies, but the lithium-ion battery is the current leading contender. Unfortunately, the United States does not have a good supply of lithium[21] (table 3.5). Most of the lithium in the world is in South America and China. A report by William Tahil of Meridian International Research estimates each kilowatt-hour of battery storage capacity requires 0.3 kg of lithium.[22] The Chevrolet Volt's 16 kWh battery requires 4.8 kg of lithium. Replacing 5% of the 251 million vehicles on the road each year with PEVs would require 60,000 metric tons of lithium. This is about 3 times current world production of lithium and 1.5 times total US reserves. Replacing all of our cars with PEVs would require 1.2 MMt of lithium, roughly 10% of world reserves. That is just for electric cars in the United States; supplying lithium for electric cars worldwide would be difficult.

TABLE 3.5 *Sources of Lithium by Country (metric tons)*

Country	2009 production	Reserves
United States	Withheld	38,000
Argentina	2,200	800,000
Australia	4,400	580,000
Bolivia	N/A	9,000,000
Brazil	110	190,000
Canada	480	180,000
Chile	7,400	7,500,000
China	2,300	540,000
Portugal	490	N/A
Zimbabwe	350	23,000
Total	17,730 (excluding USA)	18,851,000

Sources: US Geological Survey, "Mineral Commodity Summary for Lithium 2010," http://minerals.usgs.gov/minerals/pubs/commodity/lithium/mcs-2010-lithi.pdf; William Tahil, "The Trouble with Lithium: Implications of Future PHEV Production for Lithium Demand," Meridian International Research, December 2006, www.evworld.com/library/lithium_shortage.pdf.

Comparison

When operating on the battery, the PEV achieves the goal of not burning fossil fuel and reaps the benefits of electric propulsion, no losses at idle, regenerative braking, and efficient electric drive. However, the battery imposes a significant penalty in weight and space. Comparing the Chevrolet Volt and the otherwise similar internal combustion engine Chevrolet Cobalt, figure 3.5 illustrates the extent of the penalty.

Note that the Volt, with its electric motor and battery weighs 654 pounds more than the Cobalt; the lithium-ion battery itself adds 375 pounds. The Volt has seating for one less passenger than the Cobalt and 3.3 cubic feet less cargo space. The high additional weight of the Volt, almost twice the weight of the battery, indicates that the Volt requires a lot of peripheral equipment in addition to the large battery. Overall, the penalty is 650 pounds and a loss of a quarter of the payload volume. That is a high price to pay for an all-electric range of 40 miles.

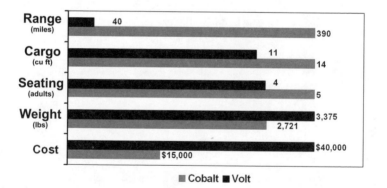

FIGURE 3.5 Chevrolet Gasoline Cobalt versus Chevrolet Volt Electric Vehicle. Though the gasoline-powered Cobalt and the electric-powered Volt when operating solely on its battery are otherwise similar, the Volt costs much more, weighs more, has less passenger capacity, less cargo space, and one-tenth the range.

Sources: Cobalt: Chevrolet, "Cobalt," www.chevrolet.com/cobalt/. Volt: Chevrolet, "2011 Volt," www.chevrolet.com/volt/.

Bottom Line

On the bright side, electric cars promise pollution-free operation and a path to weaning the automobile industry away from gasoline and oil. Unless one is satisfied with about 40 miles of driving per day and lives in a home with access to a charging circuit, the PEV will not be practical until the fast charging station infrastructure becomes as widely available as gasoline filling stations are today. This will take time. It will also be limited to availability of high-power charging sources. This will limit practicality of PEVs in remote locations.

Currently, if you want to drive your vehicle away from developed roads and filling stations, you simply take along extra cans of gasoline or have a drum of fuel delivered periodically. There is no practical way to carry extra electricity or deliver it in batch quantities.

Range on battery power is limited. Battery operation will be limited until batteries become much smaller and lighter. Meanwhile, some form of auxiliary power unit or genset will be required to provide electrical power, quite possibly for decades.

Limited availability of home charging outlets and long charging time for the outlets that are available will limit widespread deployment of PEVs until the power grid infrastructure expands. Extensive deployment of PEVs will require doubled generator capacity, an extensively expanded transmission line network, and widespread fast charging stations. Charging stations will have to be more widespread than existing gasoline stations to compensate for the shorter range and hence more frequent refueling of PEVs.

Doubling electricity demand will produce a huge increase in coal usage and corresponding pollution, and the public will not accept PEVs on a wide scale until a clean source of electricity is developed.

Low temperature will markedly reduce PEV performance in cold climates because keeping the battery warm will consume energy from the battery and adversely affect driving range.

Finally, there is a risk that the United States will become dependent on foreign sources of lithium for PEV batteries.

H-PEVs, such as the Chevrolet Volt, are neither fish nor fowl. Similar to the HEV, which has two distinct propulsion systems, the H-PEV has two distinct power sources, an externally charged battery and a fossil-fuel-burning genset, although it has only one propulsion motor. Indeed, the Volt is more of a fossil-fuel-burning vehicle than an externally charged EV. When battery and fuel tank are both full, the Volt has a range of 40 miles on battery power and 260 miles on fuel. Anyone commuting more than 20 miles one way to work—as do 23% of commuters in the United States—would be driving on fuel at least part of the time.[23] Roughly 8% of commuters would be driving on fuel half of the time. And when driving to visit my in-laws, 450 miles away, I would be driving on fuel for over 90% of the trip. Why not simplify the vehicle by running it on fuel all the time? That is, consider the SEV, depicted in figure 3.1c. The SEV is similar to the H-PEV, but it requires no external charging. All electricity is supplied by an onboard fuel-burning genset. The battery can be smaller because the vehicle seldom operates without the genset producing electricity.

The SEV would provide most of the benefits of electric propulsion (i.e., efficiency, regenerative braking, no wasted power while idling) and does not have the range limitations of P-PEVs. Since the Chevrolet Volt is advertised as getting 50 mpg on fuel, an SEV should do at least as well, probably better because the battery would be smaller and lighter. Moreover, range would not be an issue for SEVs. On the downside, the vehicle would still be burning fuel. However, different gensets could be available for different fuels. For example, if algal biodiesel becomes practical, it could be used in a diesel genset on an SEV. Perhaps most important, the SEV could be a 50 mpg or better vehicle that does not require a power grid.

A functional SEV could be deployed today and would provide greatly improved fuel economy without needing battery development or increased charging infrastructure. The SEV could then be smoothly phased out in favor of PEVs as batteries and charging infrastructure are developed.

Hydrogen Fuel-Cell Vehicles

The SEV generates electric power using an onboard fuel-burning genset. An alternative source of electricity would be a hydrogen fuel cell. An HFCV is an SEV that generates its electricity using an onboard hydrogen-consuming fuel cell (fig. 3.3b). The fuel cell generates electricity onboard simply by passing hydrogen through the fuel cell; hydrogen gas goes in, and electricity and water come out. An onboard storage tank supplies the hydrogen. One advantage of the HFCV over the P-PEV is that the battery can be much smaller, lighter, and less expensive because it is not the primary supplier of electricity for the motor, but this advantage is offset by the weight of the fuel cell and hydrogen storage tanks. Like the P-PEV, the HFCV produces no pollution or greenhouse gases. Except for the fact that the HFCV is fueled with hydrogen and not charged with electricity from the power grid, it can be thought of as a PEV with a hydrogen tank, fuel cell, and small battery replacing the large storage battery. Three advanced prototype passenger cars powered by hydrogen fuel cells were operating in 2010 (the Honda FCX Clarity, Chevrolet Equinox, and Toyota FCHV). Commercial availability of fuel cells is a decade away, but the technology has great promise.

Hydrogen Gas Supply

Onboard hydrogen storage is the first engineering issue with the HFCV. Hydrogen, like natural gas, has very little energy per cubic foot at normal temperature and pressure. The energy content of hydrogen gas at atmospheric pressure is only a quarter that of natural gas by volume (table 3.2). Onboard storage of hydrogen is more of a problem than with natural gas, though the sources of the problem are the same. As with natural gas, one could store hydrogen as a liquid, but storing it as a compressed gas would be preferred. At 5,000 psi, the pressurization of some of HFCVs, 1 gge of hydrogen takes up 9 gallons of volume. At 10,000 psi, 1 gge takes up 5 gallons. That is, even at 10,000 psi, hydrogen takes up five times the volume that gasoline would require for an equivalent amount of energy, and that is just for the gas itself. The storage tank adds weight and volume. The space penalty from hydrogen is more severe than with natural gas. Hydrogen is less practical for small passenger cars than natural gas simply because of the onboard storage constraint. One prototype, the Honda FCX Clarity, gets a range of 280 miles from a fuel tank almost the size of a 42-gallon oil barrel. By comparison, a gasoline car getting 30 mpg would have a range of over 1,100 miles with a gasoline tank this size.

The second issue is filling the hydrogen tank. More precisely, the issue is getting the hydrogen gas to the fueling station. Curiously, how hydrogen would be produced for the automotive market would depend on how the hydrogen is transported. As I said earlier, the energy content of hydrogen gas at atmospheric pressure is only a quarter that of natural gas by volume. Trucking hydrogen as compressed gas is feasible but prohibitively expensive for distances greater than 200 miles. Long-distance transport of liquid hydrogen via cryogenic pipeline would be preferable but also expensive. Moreover, the current infrastructure consists of only 700 miles of hydrogen pipeline, compared with 1 million miles of existing gasoline pipelines. The complexity of hydrogen transportation compared with natural gas and the need for massive hydrogen pipeline development probably mean that generating hydrogen at centralized production plants and transporting it to refueling stations is impractical. A better approach would be to produce hydrogen at each refueling station.

The next issue is how to generate the hydrogen. There are essentially two methods for manufacturing hydrogen: natural gas reforming and electrolysis.[24] Natural gas reforming currently produces almost all of the hydrogen used in the United States, but it seems unlikely that this method would be used for vehicle fuel. An argument in favor of reforming natural gas is that, if natural gas infrastructure were expanded to support NGVs, additional pipelines to support hydrogen production at fueling stations would not be needed. The increased natural gas infrastructure could serve both NGVs and HFCVs. While this sounds good at first, my concern is the efficient use of resources. Because of losses in the conversion sequence—natural gas to hydrogen to electricity—much more natural gas would be consumed per mile of driving than if the natural gas were used directly in NGVs. Using natural gas in NGVs would be a more efficient usage of natural resources and would generate less pollution and greenhouse gases than reforming it into hydrogen.

The most practical procedure for manufacturing hydrogen for HFCV would be to make hydrogen at the fueling station by electrolysis, passing electricity through water to make hydrogen and oxygen. If we expand the national electric infrastructure to support increasing demand and PEVs, no additional infrastructure development would be needed to support HFCV fueled by local electrolysis. Still, the scope of electric power grid expansion would increase, and the need for clean sources of electricity would increase significantly.

Comparison

While it is true that hydrogen weighs less than gasoline for equivalent energy content, the situation changes dramatically when the storage tank is considered. A gallon of gasoline weighs 2.8 kg; 1 gge of hydrogen weighs 1 kg. Compressed to 10,000 psi, 1 gge of hydrogen takes up 5 gallons. Current storage tank technology provides 6% of total weight as hydrogen. That is, 2.8 kg of compressed hydrogen in a tank weighs 17 kg. Hydrogen at 10,000 psi and the tank together weigh about six times the equivalent amount of gasoline and take up about six times the volume. The six-to-one weight and volume penalties are obvious when we compare similar vehicles, such as the Chevrolet Equinox gasoline and HFCV models (fig. 3.6).

The fuel system in the HFCV stores 4.2 kg of hydrogen at 10,000 psi weighs 300 pounds and takes up 42 gallons of space. The HFCV weighs quite a bit more than the conventional-engine car, even though the weight is minimized by aluminum doors and a carbon-fiber hood.

Hydrogen fuel-cell technology exacts a significant penalty in weight and payload space, both cargo space and passenger capacity. Indeed, the weight penalty is more severe than one might expect just from the weight and volume of the fuel-cell technology. The penalty is more than 800 pounds of added weight, loss of space for one passenger and almost half the cargo volume, and half the driving range. While the penalty is severe and driving range is less than desired, the driving range of the HFCV is greater than that of P-PEVs. Moreover, the temperature operating range of current hydrogen fuel cells is –13°F to 113°F, which makes the use of HFCVs in extremely hot or cold climates impractical.

Figure 3.7 compares energy efficiencies of P-PEVs and HFCVs by illustrating how effectively electric power from the power grid is used. Since consuming natural resources ultimately generates the power in the power grid, this figure compares how efficiently the two technologies use natural resources. The P-PEV battery is charged directly from the power grid. The battery charger converts alternating current (AC) from the power grid to direct current (DC) and stores the energy in the battery. Battery chargers are typically 85% efficient. Fueling the HFCV is more complicated. To generate hydrogen, the power supply converts AC from the power grid to DC, which powers the electrolyzer that manufactures hydrogen. The hydrogen is compressed and fed into the storage tank. Hydrogen flows from the storage tank into the fuel cell to generate electricity to power the car. In some cases, the power from the fuel cell goes directly to the propulsion motor; in other cases it goes to charge the battery. The corresponding end-to-end efficiencies dif-

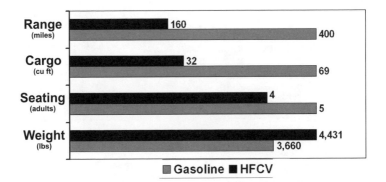

	Range (miles)	Cargo (cu ft)	Seating (adults)	Weight (lbs)

Range (miles): HFCV 160, Gasoline 400
Cargo (cu ft): HFCV 32, Gasoline 69
Seating (adults): HFCV 4, Gasoline 5
Weight (lbs): HFCV 4,431, Gasoline 3,660

■ Gasoline ■ HFCV

FIGURE 3.6   Chevrolet Equinox Gasoline Vehicle versus Equinox Hydrogen Fuel-Cell Vehicle. Although the Chevrolet Equinox gasoline and hydrogen fuel-cell models are otherwise similar, the fuel-cell car weighs more and has less passenger capacity, half the cargo space, and less than half the range. Cost is not compared, as the fuel-cell Equinox is not available for sale.

Sources: Gasoline model: Chevrolet, "2011 Equinox," www.chevrolet.com/equinox/. Fuel-cell model: Consumer Guide Automotive, "2009 Chevrolet Equinox Fuel Cell: Overview," http://consumerguideauto.howstuffworks.com/2009-chevrolet-equinox-fuel-cell.htm.

fer, but both are nominally 30%. The HFCV's efficiency is about one-third that of the battery-powered P-PEV, a comparatively inefficient use of natural resources. The benefit of the HFCV relative to P-PEV is unclear. On the one hand, the HFCV offers greater range and the possibility of roadside refueling from trucks carrying high-pressure hydrogen tanks. On the other hand, it uses natural resources less efficiently.

Bottom Line

Fuel cells are too expensive now to be practical, although prices are declining steadily.[25] The easiest and most economical method for providing hydrogen at fueling stations is to generate hydrogen by electrolysis at the fueling station. New hydrogen transportation infrastructure would not be needed because hydrogen fueling would take advantage of the improved electricity infrastructure that

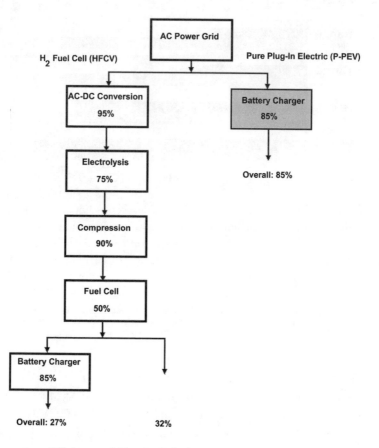

FIGURE 3.7 Efficiency of Plug-In Vehicles versus Fuel-Cell Vehicles. This chart compares energy efficiencies of P-PEVs and HFCVs by examining how effectively electric power from the power grid is used. The P-PEV battery charger stores the energy in the battery, typically at 85% efficiency. The HFCV power supply converts AC from the power grid to DC to power the electrolyzer, which produces hydrogen, which in turn is compressed and fed into the storage tank and then flows into the fuel cell to generate electricity. In some cases, the power from the fuel cell may go directly to the propulsion motor (for 27% overall efficiency) or may be stored in a battery (32% overall efficiency).

Source: Data from Ulf Bossel, "Does a Hydrogen Economy Make Sense?" Proceedings of the IEEE 94, no. 10 (October 2006).

would be required by PEVs, and natural gas would not be diverted from direct use as a fuel. However, deployment of HFCVs would increase the need for a clean and abundant source of electricity.

The fuel cell and 10,000 psi hydrogen tanks and peripheral equipment impose a significant weight and payload volume penalty on the HFCV. The range of HFCVs currently falls midway between the 40-mile range of the H-PEV Chevrolet Volt when operating on battery alone and the 244-mile range of the Tesla Roadster.

The fundamental trade-off is the more efficient use of resources by the P-PEV for the greater range of the HFCV. Moreover, even though transporting large quantities of hydrogen by truck is impractical, roadside refueling of HFCV is possible, whereas such roadside assistance to P-PEVs is not.

Summary

The goals of a new energy strategy are to reduce gasoline consumption, reduce pollution caused by burning fossil fuels, and reduce dependence on foreign oil. How well do the alternative vehicles stack up against these goals? What should the strategy be for transitioning to alternative vehicles? And what are the implications for the national energy strategy? I believe that the data speak loudly and that the conclusions I present here are justified. However, unexpected developments are possible; some promising projects will fail, while some seemingly dubious projects will succeed. In the end, I am presenting my personal opinion of what the data mean.

The first thing we should do is stretch our gasoline supplies as much as possible. Improving the standard internal combustion engine is the simplest step. The newer gasoline cars are getting better and better fuel economy, and meeting the new CAFE standard will be a good first step. However, there is not much room for additional improvements in fuel economy through engineering. We have gone about as far as possible. The same is true of diesel cars. Moreover, while diesel fuel economy, in miles per gallon of diesel fuel, is superior to fuel economy in gasoline cars, measured in barrels of oil per mile, diesel cars are less efficient than gasoline cars. Allowing the percentage of cars using diesel engines to grow too much could increase oil consumption. The strategy should be to aim for the optimal balance of the two engines.

NGVs do not pose any technological uncertainty, and they would simultaneously reduce pollution and greenhouse gas emissions and reduce consumption of gasoline and diesel. However, a natural gas pipeline network is needed to support fueling stations. In addition, the need for compressed gas storage cylinders adds a weight and cargo space penalty, which becomes more and more severe as cars become smaller and lighter (and more fuel-efficient). A corollary of the cargo volume penalty is that the fuel tank capacity is reduced so that the driving range of an NGV is less than the range of a comparable gasoline car. The final consideration is that domestic natural gas supplies are limited, and we are already importing a significant amount of natural gas. Though extensive deployment of NGVs would reduce consumption of gasoline, it would simply exchange dependence on foreign oil for dependence on foreign natural gas. While NGVs can play an important role in reducing noxious emissions and reducing demand for oil, the optimum strategy would be to use natural gas on depot-fueled fleets of larger vehicles like delivery vans and buses. Refueling at depots would minimize the number of fueling stations and hence minimize the need for infrastructure development, and placing natural gas engines on larger vehicles would minimize the size and cargo volume penalty.

Flex-fuel vehicles, those that burn E85, ethanol mixed with gasoline, definitely extend our supply of gasoline. E85 is much more expensive per mile than gasoline, making it unattractive now. This could change in the future. Even so, driving range is much shorter with E85 than with gasoline. As long as gasoline is available, many drivers would simply continue to use gasoline. We will need to keep gasoline on the market or convert all of the small planes, boats, lawn tools, all-terrain vehicles, home generators, and other small engines to E85 as well. Moreover, growing corn for ethanol takes cropland out of food cultivation. With limited cropland, we would soon become dependent on foreign sources of ethanol, and the effect on world starvation would be catastrophic. Ethanol supplies will probably be quite limited. Moreover, the negative aspects of ethanol's corrosiveness will limit widespread use of E85. Nonetheless, ethanol has some benefits in limited situations. Flex-fuel vehicles could extend our gasoline supply but only to a relatively small extent.

Biodiesel does not require engine modification, so one of my concerns with flex-fuel vehicles is not an issue for biodiesel and diesel engines. However, yields of biodiesel from crops or cooking oil are so low that they can never provide a significant percentage of US fuel demand, and crop-based biodiesel competes with food crops for arable land. We might as well take advantage of biodiesel fuel, but we must recognize that it will have only a minor effect on national demand for oil.

Biofuel, either ethanol or biodiesel, from algae might be a game changer. Demonstration projects have produced both ethanol and biodiesel from algae. Yields per acre are much higher than yields from crops, and land does not have to be taken out of food production. Successful development of algae as a source of energy would resolve many of the issues associated with biofuel.

Whatever fossil fuel is used—gasoline, diesel, or natural gas— HEV technology, as exemplified by the Toyota Prius, definitely improves fuel efficiency. Even so, while it reduces demand for gasoline, it does not eliminate it. Moreover, it involves complex and expensive machinery, having two propulsion systems, one gasoline and one electric. The HEV is probably the best thing we have today, but in the long run it has to be replaced by something simpler and less expensive.

The SEV is such an alternative. The SEV has a single, electric propulsion system powered by an onboard generator. It provides all the advantages of electric propulsion, it is simpler than the HEV, it does not require any infrastructure development, and it should provide at least the 50 mpg advertised for the Chevrolet Volt. The battery is smaller and lighter than required by a P-PEV, since the SEV almost never operates without the generator, and the motor driving the generator can be small and optimized for constant-speed operation.

There is another benefit from electric-propulsion vehicles. The distribution of energy losses to the various drivetrain components is quite different from that of an ICEV. This means that there is added opportunity for improving the fuel efficiency of conventional cars by addressing size, weight, drag, and tires. Fuel economy better than the Volt's 50 mpg should be possible for conventional automobiles, but they will have to become smaller and lighter, and the safety of mixing small vehicles with behemoths on the highway will be a major issue.

The next step would be deploying PEVs. P-PEVs, like the Tesla Roadster, are powered solely by an externally charged battery. H-PEVs, like the Chevrolet Volt, are powered by an externally charged battery but also have a fuel-burning genset to provide electric power when the battery runs low. The pure electric car offers several advantages, not the least of which is complete elimination of demand for gasoline. Problems include the cost and size of the battery, what to do when the battery runs dry, and battery charging. Battery research should bring the size, weight, and price of batteries down. A network of fast charging stations would offset the concerns about running the battery dry, but charging the battery would remain problematic. It takes a long time to charge the battery unless a fast charging station with very large power capacity is available. Home charging is severely limited by the need for off-street parking and the hours required to charge the battery. Widespread deployment of P-PEVs will have to wait for a national network of fast charging stations.

The H-PEV, exemplified by the Chevrolet Volt, solves the problem of limited charging capability by putting a genset on the car. The Volt is similar to an SEV, but the Volt relies primarily on battery operation, and its genset is primarily for backup. Making the genset the sole source of electricity, with no provision for external charging capability, as in the SEV, changes the emphasis and would allow EV technology to be optimized. In particular, the battery could be much smaller and lighter.

Range and charging time will continue to be problems. It is doubtful that batteries will improve to the point that range comparable to what we now expect from our cars will be possible with practical batteries. We will have to get used to more frequent fueling stops, and more frequent fueling will require more fast charging stations than the current number of gasoline stations, which in turn would entail a massive infrastructure expansion. Moreover, the demand for electricity will increase significantly; increased demand from EVs, population growth, and the growth of electronic gadgets will double or triple demand in a few decades. Meeting that demand will require a massive increase in electricity-generating capacity. Even with fast charging stations, the charging time would be constrained by the maximum current the battery can accept without damage. It is doubtful that PEVs will ever be charged as fast as gasoline cars are fueled.

The HFCV is an SEV, but instead of a fossil-fuel-burning genset, it uses a hydrogen-gas-consuming fuel cell to generate electricity. Although they are too expensive now to be practical, fuel-cell prices are on a track that should make HFCVs economically feasible in the future. The problem is that the large, heavy storage tanks limit range and cargo space. It is doubtful that HFCVs will be able to provide the range we are accustomed to. The main issue is supplying hydrogen at fueling stations. Manufacturing hydrogen at large plants and distributing it throughout the country is one option. However, this is expensive and requires developing a hydrogen delivery infrastructure. Although this may be feasible, two other options are more efficient. One is to distribute natural gas to fueling stations and manufacture hydrogen from the natural gas at each fueling station. Pipeline expansion would be minimal as long as we are developing wider natural gas distribution to support NGVs. However, energy losses in the hydrogen manufacturing process would mean that we would get fewer miles per cubic foot of hydrogen in an HFCV than if we used the natural gas directly in an NGV. The final option is to generate hydrogen at the fueling stations from electricity through hydrolysis. This, too, would not require additional expansion of the infrastructure, assuming that the power grid is expanded to accommodate PEVs. However, losses in hydrogen manufacture mean that the HFCV would need much more electricity from the power grid than a PEV for the same range. Consequently, the trade-off between HFCVs and P-PEVs would seem to be possibly greater range for HFCVs against greater utilization of power grid energy for P-PEVs. Additionally, emergency roadside refueling of HFCVs might be possible with service vehicles carrying high-pressure tanks of hydrogen, whereas roadside refueling of P-PEVs is problematic.

SEVs could provide excellent fuel economy. If algal biofuel becomes a possibility, SEVs might provide transportation with all the features and benefits of current gasoline automobiles but with no consumption of oil and almost no pollution. If plentiful biofuel does not become reality, electric cars, either P-PEVs or HFCVs, would provide resource-efficient, low-pollution transportation with somewhat reduced convenience, in particular in the areas of range and off-highway operation.

CHAPTER 4

Green Energy Sources

Demand for electricity is exploding with growing population and increasing demand from computers and electronic equipment. Moreover, electricity is a primary candidate for alternative automobile technology whether the emphasis is on plug-in electric cars or hydrogen fuel cells, as electrolysis is the leading contender for manufacturing hydrogen for vehicles. The unavoidable losses inherent in operating electrical power plant turbines almost triple the corresponding increase in demand for natural resources. Coal is a major source of electricity in the United States, and coal is excessively polluting and damaging to the environment. On the one hand, we need to triple our supply of electricity. On the other hand, we need to reduce pollution from coal-fired power plants. Alternative renewable sources of electricity such as solar, wind, geothermal, and hydroelectric power can play an important role in increasing our supply of electricity without increasing greenhouse gases or pollution. Perhaps renewable sources will provide enough electricity so that we can decommission some coal-fired plants and reduce greenhouse gas emissions. Even if this is not possible, we may be able to reduce noxious emissions from coal by developing clean coal.

Solar power and wind power are the primary candidates for renewable resource development. Part of the following discussion deals with how much energy is available from these sources. This is straightforward because we understand the physics and engineering issues quite well. Part of the discussion deals with the engineering aspects of getting dependable energy to the consumer, and here there are some very challenging engineering issues.

We depend on oil for many products in addition to automobile fuel. Totally eliminating demand for gasoline and diesel will not reduce our dependence on oil unless we can find other sources of nonfuel petroleum products. Either we find nonpetroleum substitutes for these products, or we figure out how to manufacture them from nonpetroleum sources such as natural gas or coal. If we cannot achieve one of these solutions, we may be able to conserve our dwindling oil supplies for essential products by obtaining gasoline and diesel from coal. The Fischer-Tropsch process provides a method of converting coal and natural gas into products traditionally obtained by refining petroleum.

Wind Power

Wind power is free, clean, and inexhaustible.[1] The promise is great, but practical considerations limit its role in a national energy policy. Wind power, like solar power, is a gift from nature. And as with solar power, the amount of wind power that is available varies with location and with time of day and season.

Available Wind Power

Maps of measured wind power provide data concerning availability of wind power. Figure 4.1 is a map of measured wind speed throughout the United States. The map shows that the potential for wind-generated electric power is only "fair" throughout vast areas of the country. Definitions of the various wind classification levels are shown in table 4.1. There are pockets of excellent potential in West Texas, the Rockies, and small pockets in the Appalachian Mountains. These might be good sources of local electricity, but they are not large enough to provide a large percentage of our national electricity needs. Wind is stronger high above the ground

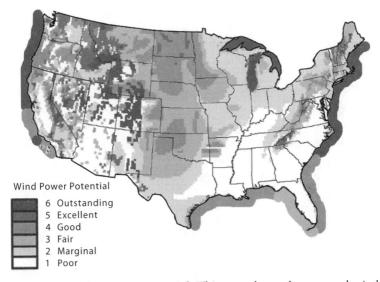

Wind Power Potential

- 6 Outstanding
- 5 Excellent
- 4 Good
- 3 Fair
- 2 Marginal
- 1 Poor

FIGURE 4.1 Wind Resource Potential. This map shows the expected wind potential throughout the contiguous United States. Table 4.1 defines wind potential classification in terms of wind speed. While there are areas of good wind onshore, good wind is limited. The best and most abundant wind is offshore.

Sources: Adapted from EIA, "Renewables," figure 13, "Wind Resource Potential," www.eia .doe.gov/cneaf/solar.renewables/ilands/fig13.html, using data from NREL, "Wind Resources and Transmission Lines," www.nrel.gov/wind/systemsintegration/images/home_usmap.jpg.

TABLE 4.1 *Wind Speed Classification*

	Wind Speed at 50 m height	
Classification	mph	m/s
Superb	19.7–24.8	8.8–11.1
Outstanding	17.9–19.7	8.0–8.8
Excellent	16.8–17.9	7.5–8.0
Good	15.7–16.8	7.0–7.5
Fair	14.3–15.7	6.4–7.0
Marginal	12.5–14.3	5.6–6.4
Poor	≤12.5	≤5.6

Source: NREL, "Wind Resources and Transmission Lines," Wind Power Classification Chart, www.nrel.gov/wind/systemsintegration/images/home_usmap.jpg.

where friction with the ground is minimal. Consequently, tall turbine towers, much taller than high-voltage transmission-line towers, are required.

The best wind, however, is offshore. Wind power potential is "outstanding" to "superb" on the Great Lakes and along the coasts, from the Canadian border to central California on the west coast and from Maine to the border between North and South Carolina on the east coast. Wind is generally stronger over water than over land because surface friction is less, and it is generally stronger along coastlines because the heating and cooling of land near water produces onshore and offshore breezes.

Wind Turbines

Many manufacturers make small, quiet wind turbines for individual homes and farms. A smaller number of companies manufacture the large wind turbines required by public power utilities (fig. 4.2). Vestas is one of the largest manufacturers of wind turbines, with 20% of worldwide market share and over thirty-eight thousand installed turbines.[2] The Vestas V90 3.0 MW turbine is a typical commercial wind turbine satisfactory for use by a public power utility. Table 4.2 summarizes the engineering specification of the V90.

Specifying the power of wind turbines and solar farms is misleading. In the case of solar farms, the specified power is available only when the sun is shining; in the case of wind turbines, the power is available only when the wind is blowing at maximum design value. In reality, both solar panels and wind turbines produce about a third of the specified power. For example, the Vestas V90 turbine is rated at 3.0 MW, but it only produces that much power when the wind is greater than 34 mph (gale force wind). According to the map of available wind power (fig. 4.1), wind this powerful is exceedingly rare. Most of the time, the wind will not be that strong, and the electrical output will be much less than the rated value. The concept of a "capacity factor" accounts for this. Common turbine capacity factors range from 0.2 to 0.3, meaning

FIGURE 4.2 Typical Large Wind Turbine. A large wind turbine in a wind farm in Virginia.

Photo by Minnie C. Gallman.

TABLE 4.2 *Wind Turbine Specifications: Vestas 3.0 MW*

Item	Specification
Rated output	3 MW
Tower height	80, 90, or 105 m
Rotor diameter	90 m
Rotation speed	16.1 rpm nominal (8.6 to 18.4)
Cut in wind speed	3.5 m/s (8 mph)
Rated wind speed	15 m/s (34 mph)
Cut out wind speed	25 m/s (56 mph)
Noise level	97.9 dB at 4 m/s (9 mph) wind;
Measured 10 m above ground on an	106.9 dB at 9 m/s (20 mph) wind
80 m tower, i.e., 70 m from the rotor	
Output	1,000 volts at 50 Hz
	≤250 kW for wind ≤5 m/s (11 mph)
	3 MW for wind ≥15 m/s (34 mph)

Source: Vestas 3.0 MW turbines, www.vestas.com/en/wind-power-plants/procurement/turbine-overview/v90-3.0-mw.aspx#/vestas-univers.

that a 3.0 MW turbine actually operates at 0.6 to 0.9 MW on average, and output at that level requires "good" wind.

How much area would wind farms cover? Assume that the wind farm uses Vestas V90 turbines. Each turbine is rated at 3.0 MW, so the average output, calculated using a 0.25 capacity factor, is 0.75 MW. The turbines should be spaced five to ten rotor diameters apart,[3] which, for the 90-meter-diameter V90 rotor, is 450 to 900 meters. When the turbines are arranged in a square array, each turbine takes up 450 × 450 to 900 × 900 square meters, 50 to 200 acres per turbine. Land requirement then is 67 to 267 acres/MW. Over the course of a year, the V90 generates 6,570 MWh (24 hours per day for 365 days), so the energy yield is 33 to 130 MWh/year/acre. The Cape Wind offshore project in Nantucket Sound expects to produce 96 MWh/acre/year, and the Horse Hollow Wind Energy Center in Texas produces 55 MWh/acre/year, both within the estimated range. Taking a middle value as a rule of thumb, one can expect 100 MWh/acre/year, or 90 acres/MW, from a wind farm in an area of good wind. This is about a third the output from a solar farm for the same footprint.

Noise from wind turbines is a concern. Data from Vestas specifications indicate a sound power level between 97.9 decibels (dB) and 106.9 dB, measured 10 meters above the ground from a V90 3.0 MW turbine, depending on wind speed. The higher number applies when the wind is greater than 9 meters per second (m/s), or 20 mph. To provide some perspective, 110 dB is a chainsaw at 3 feet; 100 dB is a jackhammer or a disco, also measured 3 feet from the source. A diesel truck with a running engine at 30 feet is 90 dB, traffic on a busy road is 80 dB 32 feet from the road, a vacuum cleaner at 3 feet is 70 dB, a busy restaurant is 60 dB, and an average home is 50 dB. Any sound level over 80 dB is potentially hazardous if exposure continues long enough. Occupational noise regulations limit exposure to 100 dB to one hour per week. Noise of this level definitely produces sleep disturbance, elevated blood pressure, and an increased risk of heart attack. The noise from the turbine would not always be above 100 dB, since the wind is not constant. However, it would be intolerable, and even unhealthy, to live close to a large wind turbine operating in wind approaching rated speed.

The farther away you are from a turbine, the less noise you experience. Adequate separation between people and wind turbines would make wind power more acceptable. Figure 4.3 shows how the noise level from a large wind turbine decreases with distance from the source. The figure depicts noise from a single 3 MW wind turbine; noise from a smaller turbine would be less, and noise from a number of turbines in a wind farm would be more. The figure clearly shows that separation of several miles is needed to ensure a noise level no louder than an average home. This is certainly possible to effect, but it limits how close to population centers large wind farms can be located to about 3 miles. There already is strong resistance to onshore wind farms near population centers. Putting wind farms offshore would alleviate the noise problem.

Placing wind farms offshore is a distinct possibility. While eight European countries have built offshore wind farms, the United States has not built even one. Some are in the planning stage. The Cape Wind project[4] is the closest to fruition. There are several dif-

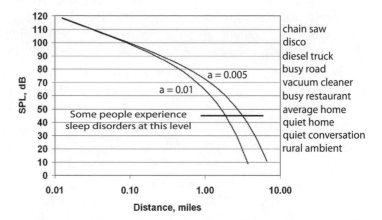

FIGURE 4.3 Wind Turbine Noise versus Distance. This figure shows the decrease in noise with distance from a wind turbine. Noise—that is, sound pressure level (SPL)—is shown on the left axis. Typical sources of noise at each level are shown on the right axis. Noise versus distance from the source is described by the equation

$$SPL = k - 20*\log_{10}(d) - a*d,$$

where SPL is the sound pressure level in dB, d is the distance in meters, a is a factor that accounts for atmospheric and ground foliage attenuation, and k is a constant. The graph shows noise versus distance for values of a typical of rural areas, and noise data from Vestas 3.0-MW turbine data (table 4.2), that is SPL = 106.9 dB at 70 meters distance and 9 meters per second (m/s) wind speed, for which the constant k = 144.5. Wind speed of 9 m/s (20 mph) is on the borderline between "outstanding" and "superb" wind and is close to the upper range of expected wind. Noise would be 10 dB less at 4 m/s (9 mph), "poor" wind. Separation between residential areas and a single wind turbine would have to be about 3 miles for the noise level to be similar to a quiet home in 20 mph wind or about 2 miles in 9 mph wind (which is not much use for generating electricity). Three miles seems to be a reasonable separation distance. Distance from a wind farm consisting of many turbines would have to be greater.

ferences between offshore and land farms. First, wind offshore is generally higher velocity and steadier than on land, so offshore wind farms should be more efficient. On the other hand, construction is more difficult, and the difficulty increases with deeper water.

One concern is potential obstruction to navigation and commercial fishing that might be caused by offshore wind farms. The Cape Wind project is located away from shipping lanes, and the turbines are widely separated. They form a grid, with 0.34 nautical miles (630 m) between turbines along each row and 0.54 nautical miles (1,000 m) between rows. This should provide adequate clearance. The wind farm will be located on Horseshoe Shoal, where the water depth can be as little as 2 feet at low tide, making this a good area for boaters to avoid. Certainly major shipping avoids the shoal. Moreover, the large number of offshore oil rigs in the Gulf of Mexico (about thirty-eight hundred) does not seem to cause undue problems for boaters.

Another consideration with offshore wind farms is visual pollution. The turbines are rather large. The hub is 258 feet above the water, and the rotor diameter is 365 feet. The concern is that these large structures will be an eyesore to people on shore.[5] Offshore visual pollution, however, seems a small price to pay for clean energy. After all, generations of urban folk have grown up without seeing a single star at night, and most of us have accepted landscapes peppered with utility poles, cell phone towers, security lights, and so on. Given the alternative of running out of energy, I believe that offshore wind turbines will eventually be accepted.

One of the overlooked problems with wind and solar power is the need for transmission lines and the public's resistance to building new ones. Stories about utility companies not being able to meet customer demands because of resistance to new transmission lines abound. T. Boone Pickens aborted his plans to build a wind farm in West Texas because he could not get financing for the necessary transmission lines.[6] Investor concern about public resistance and the time and effort needed to bring the project to fruition were undoubtedly major factors.

The Pickens Plan sets a goal of producing 22% of our electric energy from wind.[7] At 100 MWh/acre/year, roughly 14,000 square miles of wind farms would be needed to meet 22% of our current national demand for electricity. Finding this amount of adequate wind onshore (Pickens planned on an onshore location in West Texas) does not seem to be feasible. If the wind is farmed offshore, it would require a band of wind turbines 8 miles wide running from Maine to the North Carolina–South Carolina border (1,000 miles) and from the Canadian border to San Francisco (700 miles). That is a lot of coastal area, and wind-farming such an area is probably impractical. If as much as 20% of the coastal band were used, wind power could not provide more than 4.4% of the current national demand for electricity.

Wind Dropouts

A major concern about wind power is that it is intermittent. Wind varies in velocity and occasionally dies completely. Unfortunately, these wind dropouts are sporadic and unpredictable. No electricity is generated when there is no wind, which poses two concerns. The first is that if a wind farm is the primary source of power for a community, wind dropouts would present a serious problem for the community unless there is some backup power source. The second concern is that if a wind farm connects to the national power grid, wind dropouts could affect grid stability. In February 2008, a wind farm in West Texas almost brought the national grid down when the winds died.[8] The wind died for three hours, and generated power fell 75%, dropping 1,500 MW, just as there was an evening spike in demand. The Texas grid nearly collapsed. I discuss grid stability more thoroughly in the section dealing with solar power, but the essential point is that the national power grid cannot handle such large unexpected variations.

Providing a steady level of power twenty-four hours a day and protecting the power grid against intermittent supply outages requires some form of energy storage at each wind farm. A big storage battery sounds impractical at first, but Tokyo Electric Power Company of Japan pioneered the use of sodium-sulfur (NaS) batteries two decades ago.[9] Tokyo Electric Power Company has been deploying them with Japanese wind farms, and American Electric Power, headquartered in Columbus, Ohio, is working on similar systems in this country. The NaS battery has the highest energy density of any battery and nearly 90% efficiency, making it a good choice for power backup. However, the battery operates at 290°C to 360°C (550°F to 680°F) and contains very caustic materials, making it impractical for mobile or home use. A NaS energy storage system capable of providing about 1 MW of power for seven hours weighs about 88 short tons and fills a space the size of two semitrailers. The cost, at present, is several million dollars.

Bottom Line

Wind power is definitely an option, but the amount of energy wind power can provide is limited. Public utility wind farms can provide about 100 MWh/acre/year. About 14,000 square miles (an area about the size of Maryland) of wind farm would be required to supply 22% of current US demand. To generate the same amount of electricity offshore would require a band of turbines 8 miles wide along 20% of both coasts. However, wind power will most likely not be a major contributor to national power supply, probably no more than 5%.

Onshore areas of adequate wind are quite limited. Noise pollution and limited availability of good wind onshore will probably restrict most public utility wind farms to the Great Lakes and offshore, where construction costs could be high. Offshore wind farms should not be navigation hazards, since they would be in relatively shallow water whenever possible, and the hundreds of offshore oil platforms in the Gulf of Mexico do not seem to bother many people. The visual pollution of near-coastal installations is something that we might have to accept, just as we have gotten used to cell phone towers and utility poles and lines.

Though not as sporadic as solar power, wind does occasionally die. An energy storage system such as a NaS battery is almost essential for commercial installations. Careful site selection for steady wind and battery backup should minimize loss of power during sporadic short periods of no wind and unacceptable short-term variation in power output.

Solar Power

Solar power is free, clean, and inexhaustible. The promise is great, but practical considerations limit its role in a national energy policy.

Available Solar Power

The first thing to look at is the amount of solar power available from nature. The numbers and the physics are well known. Solar irradiance averages 1,370 watts of power over each square meter (W/m^2) at the top of the atmosphere. There is some variation over the decades because of fluctuating solar activity, there is an annual cyclic variation caused by seasonal changes in the distance from the sun to the earth, and different instrument systems yield slightly different values. Still, 1,370 W/m^2 is a good value for planning purposes. Approximately half of the energy at the top of the atmosphere gets through to the earth's surface directly below the sun. The rest is lost to absorption and reflection caused by clouds, dust, and pollution. Less energy gets through at higher latitudes because the path through the atmosphere is longer than at the equator. Available power varies during the day and as the seasons change. It reaches its peak value at noon, is nonexistent at night, and is greater in summer, when the sun is more directly overhead. A working figure for *average* solar power over the course of a year in the United States is 228 W/m^2, so the amount of solar *energy* available per day (average power multiplied by 24 hours) is 5.5 kWh/m^2. More power is available at noon in the summer, none is available at night, more is available in areas of the country blessed with clear skies, and less is available in more frequently overcast and rainy areas.

Measured data provides more detail. The map in figure 4.4 shows insolation (incident solar radiation) ranging from 4 to 5 kWh/m^2 per day in the Northeast to 6 to 7 kWh/m^2 per day in the Southwest, corresponding to power averages over an entire year ranging from 167 to 208 W/m^2 in the Northeast to 250 to 292 W/m^2 in the Southwest. These numbers assume the collector is perpendicular to the sun's rays.

The amount of sunlight intercepted depends on the effective

Tilt = Latitude
kWh/m^2/day
8-9
7-8
6-7
5-6
4-5
3-4
2-3

FIGURE 4.4 Solar Photovoltaic Resource Potential. This map shows the available insolation across the contiguous United States averaged over a year. That is, the power available at any specific time varies with season and time of day. In particular, no solar power is available at night or during periods of overcast. The map shows clearly that almost twice as much solar energy is available in the Southwest as in the Northeast. Putting solar farms in the Southwest would maximize electrical output but would require large transmission lines to distribute the electricity throughout the country.

Source: Adapted from EIA, "Renewables," figure 11, "Solar Photoltaic (PV) Resource Potential," www.eia.doe.gov/cneaf/solar.renewables/ilands/fig11.html.

collector area—that is, the actual area of the collector multiplied by the cosine of its tilt angle relative to the sun's rays. A collector flat on the ground at my home in Virginia (latitude 38° N) receives 79% of the power that would be collected if the panel were tilted correctly. To overcome these losses, many collection systems track the sun during the day and seasons, keeping the flat collector panel perpendicular to the sun's rays (fig. 4.5). The more sophisticated tracking systems collect more power but with increased system complexity and cost. Details for a variety of collector configura-

FIGURE 4.5 Nellis Air Force Base Solar Power Farm. This photograph shows a solar farm made up of individual sun-tracking collectors, each comprising twelve individual PV panels. Space for access between collector units and for support buildings such as those seen in the background here reduce the amount of solar power that can be captured within a given solar power farm footprint area.

Source: Nellis Air Force Base, www.nellis.af.mil/photos/media_search.asp?q=solar array& page=2.

tions are available from the National Renewable Energy Laboratory (NREL).[10] The NREL website breaks down results by month and by different collector alignments.

Figure 4.5 shows the flat photovoltaic (PV) panels of the Nellis Air Force Base array.[11] This large array automatically tracks the sun during the day and keeps the panels perpendicular to the rays of the sun, maximizing electrical output. One thing to notice is the space between the individual units required for access, roads, and so on. The picture also provides a feeling of how much land area is needed for a commercial solar plant.

A fixed collector in the United States tilted perpendicular to the sun's rays receives 228 W/m^2/day (power averaged over a day), which is 5.5 kWh/m^2/day (total energy collected over a full day). Both figures are averages over a year and averaged over the whole country. That is, a collector receives more energy in summer than in winter, and a collector in the Southwest gets 50% more power from the sun than one in the Northeast. Continuously tracking the sun increases power about 50% over a fixed collector. The type of tracking and the location of the collector are very important. A commercial station in the Southwest that uses tracking receives over twice as much solar power as a fixed collector in the Northeast.

How much of the incident solar power becomes useful electricity? This can be determined by examining each stage in the process of converting sunlight to useful electricity. I use 15% for the efficiency of photovoltaic cells, which is better than existing commercial photovoltaic cells but is plausible with ongoing research and development (R&D). The PV cell feeds an inverter that converts the direct current (DC) from the cell into alternating current (AC) for the homeowner or power grid. Inverter efficiency is typically 85%. If we factor in a 12% loss from dirt, collector misalignment, and component aging, overall system efficiency at converting available solar power to AC electricity comes to about 10%. Consequently, the engineering limitations reduce the US average annual input from a fixed collector from 228 W/m^2/day to 23 W/m^2/day, or 0.55 kWh/m^2/day of usable AC electricity. This is an average figure for the United States, and it assumes a fixed collector. If the collector is located in the Southwest, which has more sun, and a two-axis tracking collector is used instead of a fixed collector, usable electricity is roughly 50% greater. PV cell R&D will improve efficiency and reduce cost, but it is unreasonable to expect that overall efficiency will ever be much higher than 20%. The 10% figure is good for planning purposes. That is, the numbers for available solar power shown in figure 4.4 can be divided by 10 to determine usable solar electricity.

How do we account in our calculations for the sun's shining only a few hours a day? To estimate how much energy can be derived from a home solar panel, 0.55 kWh/m²/day of household AC power is a good planning figure. This is an average over twenty-four hours. All of the power is collected during the roughly eight hours of best daylight, but nothing is collected for the other sixteen hours.[12] To draw power continuously for a full day, one needs a battery with capacity two-thirds of the total. For the eight hours when the sun is shining, the solar panel provides electricity to the load and charges the battery; the battery provides electricity to the load for the remaining sixteen hours. The map in figure 4.4 gives insolation averaged over a full day. To draw that power continuously over a full day, one also needs a storage system capable of storing two-thirds of the total daily energy.

You also need to remember that these figures apply to *collector area*. If you are planning a small system for your home, you can use the numbers as given. However, if you are considering a large public utility solar farm, you have to recognize that the entire solar farm has to have space between collectors for access, offices, roads, and other spaces not devoted directly to photocells. The energy you can get from the entire footprint of the solar farm is less than the energy available from the collector area. You have to include the solar farm footprint factor, which I will do shortly.

Finally, one has to factor in transmission losses. Your home system provides energy right at the point of consumption, your home. A public utility will be located in an area of good sunlight, which may be far from where the electricity is used. Power is lost in the transmission lines connecting generator to consumer. These losses average 7% in the US power grid and would be greater with longer transmission lines. If we were to power the entire country from large solar farms in the Southwest, as some suggest, greater losses would be incurred because of the long transmission lines. Therefore, for a public utility, you have to derate the available energy by at least 7% for transmission line losses.

Solar Power for My Home

I set out to design a solar electric power system for my own home. First, I figured out what would be required to power a 100 W lightbulb continuously. Using the nominal 23 W/m^2 power for a solar system in my area (Virginia, just outside Washington, DC), I figured I would need slightly less than 5 square meters of collector (7 feet by 7 feet). This seemed feasible (I will get to cost later), and I moved on to powering my whole house. My monthly electricity consumption ranges from 480 kWh to 2,200 kWh depending on the season (I have electric air conditioning and gas heat). My annual consumption is 12,000 kWh, which is almost exactly the average annual US individual residential household consumption. This is an average power usage of 1,400 kW, which would require 60 square meters (650 square feet) of solar collector. Since the foundation of my house is 900 square feet, it looked as if I could power my home by covering the roof with solar panels.

Then I went looking for a system. A worksheet on the first website I came across that sells solar power systems for home use indicated that I would need a system with 520 square feet of collector that would produce 1,125 kWh/month.[13] Since this calculation was in general agreement with my theoretical estimate, I proceeded to consider how I would handle intermittent sunlight.

The numbers I have been working with are averages over several days. What do I do at night when the sun is not shining, or when clouds or rain obscures the sun? There are three options: an off-grid system, a gridded system, and net metering. In an off-grid system, my house would not be connected to a commercial power grid, and I would have no electricity unless my solar power system supplied it. I would have to include enough batteries in my system to carry me through the sunless periods. This would not be out of the question, but it would add to the cost. Lack of sun at night is predictable, and one can calculate how much battery backup is required. Periods of rain and clouds are much harder to predict and may last for days. I would have to install more batteries to carry me for several sunless days, and the cost would be still higher. Even then, if I encountered a longer-than-predicted sunless period, I would have to make do without electricity. And what would I do about the months when my air conditioning raises demand to twice the average? I would have to double the collector size to provide increased power and double the size of battery backup. At this point, size, complexity, and cost become prohibitive. An off-grid system seems practical only if you live in a sunny area with pleasant weather, or if your demand is low. It would not work for my home in suburban Virginia.

In a gridded system, the next option, my house would stay connected to the power grid. This gives me a simple strategy. I would use my solar panels when the sun is shining and purchase electricity at night or when the solar power system could not keep up with demand. Of course, by staying on the grid system, I would be adding the expense of electricity. I would still install a battery system large enough to get me through the night and a day or two of bad weather so I could minimize the amount of energy I would have to purchase. I would end up purchasing electricity in summer for air conditioning and during long periods of bad weather, and I would generate excess power in winter when my demands are low. This would be an inefficient way to provide power for my house.

The third option is net metering. Under net metering, my house would stay connected to the power grid so power would always be available, as in the gridded system. The unique feature of net metering is that the solar home system pumps excess electricity into the power grid, essentially selling unneeded power to the utility company. For much of the day I would use solar power, and at night I would purchase from the grid. When the sun is shining and I am not consuming power, I would sell excess power to the utility. I might even make a profit if my solar system generates more power than I consume. However, when I looked into net metering, I found the old adage that the devil is in the details to be true.

Net Metering in Virginia

The first annoying detail is that the program is limited to residential generator systems up to 10 kW. Since the recommended system for my house exceeds this, I would have to make do with a smaller system, which would not provide as much of my electricity using solar power but would cost less and would allow me to enroll in net metering. I might save money this way, but I would not be supplying as much of my electricity from solar power as from coal-fired power plants, which was the whole purpose of this exercise.

The second problematic detail is that enrollment is open on a first-come, first-served basis until the rated generating capacity owned and operated by customer-generators in Virginia reaches 1% of each electric distribution company's adjusted Virginia peak-load forecast for the previous year. Under these regulations, home power systems (solar, wind, or whatever) would not be allowed to supply more than 1% of the power. Thousands of home generators tied to the grid providing most of the nation's power is apparently not acceptable. Net metering is not available to very many people.

Why does the utility impose these restrictions? The answer has far-reaching implications for the future of both solar and wind power. You have to understand that the "grid" is a nationwide system of interconnected power stations, consumers, and transmission lines. What happens in Sheboygan affects power distribution in Los Angeles. Maintaining grid stability is difficult. Currently, with about 450 electric power generators throughout the United States interconnected, with connections to Mexico and Canada, keeping the system from crashing is a full-time effort. Managing the process of taking one power plant off line or bringing one back on line requires a lot of coordination around the country. Even the unexpected failure of a single relay can cause disaster, as happened in the 1965 blackout in the Northeast that disrupted power to 25 million people in Canada, New England, New York, and New Jersey. It happened again in August 2003, in the same area, affecting some 55 million people. Winds die, and turbine output drops; winds increase beyond the turbine design limits forcing the turbines to be shut down; clouds, storms, and dust pass over solar collectors, and output drops suddenly and picks up again just as suddenly. The national power grid cannot handle such large unexpected variations. The thought of thousands of home generators, solar systems, and wind turbines connected to the grid going up and going down as homeowners make adjustments or as clouds drift over the country must terrify the engineers who manage the grid. Asking the system to accept a tenfold or one-hundred-fold increase in the number of power plants, and erratic power to boot, is unreasonable. It will only be practical if the grid's stability is markedly improved.

The situation is even more complicated. When a friend of mine installed a gridded home solar system in his house in Maryland, the interface between the solar system and the power grid provided by the utility company *automatically disabled the solar power system if the grid lost power*, in order to protect the grid. That is, with a gridded solar power system a backup generator is needed when grid power is lost, even when the sun is shining brightly.

Cost

An off-grid system would not work for me because of my heavy air conditioning demand in the summer. The simple gridded system would cost over $71,000 plus installation. My total annual expense for electricity is $1,200. I might save half of this by supplying some of my electricity with a solar system. Spending over $71,000 for a solar system that saves $600 a year is not attractive to me. The cost of individual solar systems has to come down ten-fold to be widely accepted. Perhaps this is possible, but it is a long time in the future.

Commercial Solar Utilities

Commercial utility power plants, being large and expensive facilities compared to home systems, can be more complex as economies of scale come into play. They would generally use optimized tracking of the sun and would be able to capture more of the solar power. For example, using the NREL site mentioned earlier, we can calculate that a site in the southwestern United States using two-axis sun tracking could expect 0.65 kWh/m^2/day—better than average. This is equivalent to 237 kWh/m^2/year, or 960 MWh/acre/year. These numbers assume that the solar panels completely cover the ground. While this might be practical for a system that serves a single homeowner, a solar farm large enough to serve a community needs access roads, space between panels for maintenance, and administration buildings. Calculating from a survey of existing or planned solar farms that indicates a spatial utilization factor between 0.3 and 0.4, one can expect between 288 and 384 MWh/acre/year from a commercial solar farm. Taking the middle value, one arrives at a rule of thumb for solar farms of 300 MWh/acre/year of usable electric energy.

At 300 MWh/acre/year, a conservative estimate of commercial solar farm output, a typical solar farm covering 21,000 square miles could provide all the electricity consumed in the United States (4.1 trillion kWh in 2008; see table 1.3). This corresponds to an area the size of the state of West Virginia, or one-fifth the size of Nevada. The footprint of the solar farm itself is only part of the land needed. Getting the power to consumers would require new transmission lines.[14] For a simple estimate of what would be needed, assume one new transmission line from the solar farm in the Southwest to each of the lower forty-eight states. Further, assume that the distance from the Southwest to a state averages 2,000 miles. The minimum length of transmission lines is 96,000 miles. Now recognize that transmission lines do not go directly from point to point but weave around. Now we are up to 100,000 to 500,000 miles of new transmission lines. Each high-voltage transmission line requires 120 to 140 feet right of way, so new transmission lines from our hypothetical solar farm would require between 2,000 and 10,000 square miles of land.

That is not the end of the story. Energy is lost in transit. The rule of thumb is that losses in the US grid amount to a little over 7%. One drawback to putting a huge solar farm in Nevada where the sun is good is that long transmission lines are required to get the power to the rest of the country. Transmission line losses will be greater than 7%, so the farm will have to be bigger to compensate. If we assume that transmission losses reduce the useful energy we can get from a remote solar farm by about 10%, the median value for solar farm power production is reduced to about 270 MWh/ acre/year. We would then need 21,000 to 23,000 square miles of solar farm and 2,000 to 10,000 square miles of transmission line right-of-way to meet current US electricity demand.

If we look into the future and assume that demand for electricity will triple as a result of population increase, growth in demand from electronics and computers, and widespread deployment of electric cars, then we find ourselves needing about 100,000 square miles of land devoted to solar farms to provide all of US electricity demand. This is an area the size of Nevada, which is clearly not practical.

When I started developing this book, I often saw press releases and news articles extolling a new solar power plant and the thousands of homes it would supply. I was generally impressed. However, after working out the numbers, I am no longer as impressed. A solar power plant that provides all the electricity needed by 1,500 homes sounds good, but the electric energy consumed by these homes is a tiny percentage of the national consumption, 0.00036% to be exact. The electric energy consumed by a million homes is only one-third of 1% of the electric energy consumed in the United

States in one year. Providing electricity to a few thousand homes makes solar power sound good, but it is insignificant from a national perspective.

Backup for the Diurnal Cycle

We cannot ignore the mismatch between consumption of electric power in the modern world and the diurnal variation in sunshine. Back in the 1930s, our nighttime usage of electricity was a fifth of our daytime usage. Now our nighttime usage decreases not even by half. We use almost as much electric power at night as we do during the day. This actually benefits the electric power industry because it means power plants operate at almost full capacity most of the time, a very efficient way of operating. However, it means that we need the same generator capacity at night as we do during the day. Even if solar power could provide all of our demand for electricity during the day, we would not be able to reduce the number of conventional power plants because they would still be needed at night. Moreover, we would have the twice-a-day chore of bringing plants online or taking them offline.

It seems plausible that NaS batteries can smooth out the intermittent supply of wind power. Doing the same thing with intermittent solar power is more of a challenge because the predictable periods of little or no sunlight over the diurnal cycle are much longer, at least sixteen hours every day, and the unpredictable periods of inadequate sunlight because of bad weather may last several days. Battery backup is probably not feasible. However, the need for a constant source of electricity is the same, both to meet demand and to ensure grid stability. One possibility is to install a fossil-fuel generator at each solar farm to maintain constant output regardless of the ambient sunlight. This would solve the problem but would be cumbersome and an inefficient utilization of the generators.

Another possibility is to use concentrating solar power (CSP) thermal systems.[15] The discussion so far has concentrated on PV systems, which convert sunlight directly into electricity. CSP systems use sunlight as a source of heat, which in turn drives a generator. There are three basic types: linear, dish, and heliostat tower. Linear systems consist of a long trough-shaped mirror that focuses sunlight onto a tube containing a working fluid.[16] The heated fluid drives a turbine, which drives a generator. Dish systems use a parabolic dish reflector that focuses sunlight at a point.[17] The most common dish system design puts a Stirling engine generator at the focus point. The parabolic dish is inherently small, and a solar farm would consist of many small parabolic dishes, each with its own Stirling generator. A heliostat tower system uses numerous heliostat reflectors arranged to focus reflected sunlight at a single point at the top of a tower, effectively making a single, very large parabolic reflector. The focused sunlight heats a working fluid, which drives a turbine generator.

CSP systems require the same land footprint as PV systems, since they all depend on collected sunlight, and the CSP and PV systems are equally efficient in converting sunlight to electricity.[18] However, the CSP systems that heat a working fluid offer a possibility of providing power when the sun does not shine. The concept is to store heat energy in the working fluid when the sun is shining and use stored heat to drive the turbine when the sun is not shining.

Bottom Line

Solar power is a mature technology readily available to consumers as off-the-shelf products. Solar power is indispensable in situations and locations where the national power grid is not available. Nonetheless, solar power is too expensive now to be practical for individuals when connection to a commercial utility is available. PV cell R&D will bring prices down, but the tenfold improvement necessary to make residential solar power generation practical is unlikely. Net metering may assist a few people, but grid stability will limit the number of people who can take advantage of net metering, and it will have little effect at the national level.

Commercial solar farms can produce about 300 MWh/acre/

year on average. Public-utility-scale commercial systems suffer from their huge footprint, including both the solar farm itself and the land required for transmission lines. Providing 100% of the nation's current electricity demand would require 23,000 to 33,000 square miles of solar farm and transmission lines plus 3 billion MWh of energy storage. The land area is perhaps acceptable, about the size of Ohio, or 1% of the entire country, or a quarter of the National Park Service holdings. However, community resistance to new transmission lines is already very strong, and the cost of battery backup would be prohibitive. The projected threefold increase in demand for electricity in the coming years raises the total footprint to over 100,000 square miles, an area about the size of Nevada.

Though there is much talk about improving cost and efficiency of PV cells, the really important issues surrounding solar power are grid stability, providing power during dark periods, the large footprint of solar farms and transmission lines, and public resistance to these new transmission lines.

The diurnal solar cycle and bad weather mean that solar power systems experience more frequent and longer periods of dropout than wind systems, and the battery backup system that seems to be feasible for wind farms probably is not feasible for solar systems. However, CSP systems that heat a working fluid may provide constant electricity. Otherwise, it seems that a backup generator would be needed at each solar farm.

Geothermal Power

The interior of the earth is very hot, and in certain circumstances this heat is a practical energy source. Like solar power and wind power, geothermal energy is freely available, everlasting, and pollution-free. And as is the case with solar and wind power, the amount of available geothermal energy varies with location throughout the country.

There are three common uses of geothermal energy.[19] First, hot water from springs and reservoirs near the surface may be used to heat buildings directly. To be practical, a source of hot water has to be near the surface and located where the heat is needed. Usually, this means areas where hot springs, geysers, and so on are common. An extreme example is the system in Reykjavik, Iceland, which provides 95% of the city's heat. Much of Iceland sits atop an active volcanic rift zone, and Reykjavik's unique situation makes citywide heating possible. Other locations are not so fortunate.

Second, geothermal heat pumps use underground stable temperature near the surface to heat and control the temperature of buildings. For this application, there has to be a reservoir of room temperature near the surface that the heat pump can tap into and use as a heat source in winter heating and as a heat sink in summer cooling. Geothermal heat pumps use water pipes buried underground as a heat exchanger. These systems require adequate area for the heat exchanger, the ground must permit burying pipes, and the temperature must be right. Geothermal heat pumps are practical throughout the entire United States, although local conditions may make them impractical at individual locations.

The third application of geothermal energy is the one I am most interested in, generating electricity. Geothermal electricity power plants use steam or hot pressurized water from deep in the earth to drive turbines that generate electricity. The United States is the world leader in generating geothermal electricity. Even so, the amount is small. In 2009 geothermal energy provided 0.35% of the electricity consumed in the United States.[20]

There are three main types of geothermal power plants.[21] Dry steam plants use steam piped from a geothermal reservoir to drive the generator turbine (fig. 4.6). The first geothermal power plants were dry steam. These are simple and effective but require a reservoir of high-pressure steam to tap. After the steam passes through the turbine and condenses, it is pumped back into the ground, where it helps to maintain the supply of steam. One disadvantage

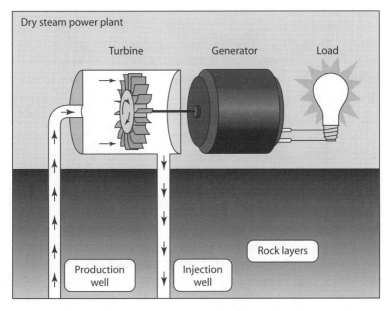

Dry steam power plant

Turbine   Generator   Load

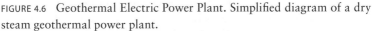

Rock layers

Production well   Injection well

FIGURE 4.6   Geothermal Electric Power Plant. Simplified diagram of a dry steam geothermal power plant.

Source: Adapted from EERE, Geothermal Technologies Program, "Hydrothermal Power Systems," www1.eere.energy.gov/geothermal/powerplants.html.

of dry steam plants is the possibility that the supply of steam may unexpectedly disappear, either temporarily or permanently.

Flash steam plants use high-pressure hot water from deep reservoirs that is "flashed" to steam that drives the generator turbine. When a high-pressure fluid sprays into a low-pressure tank, most of the fluid instantaneously vaporizes or "flashes" into steam.[22] Flash steam plants require very hot water, at least 180°C (360°F). Unfortunately, areas of geothermal temperatures this high are limited to the far western United States (fig. 4.7). Most geothermal power plants are flash steam plants.

A third type of geothermal power plant, the binary cycle plant, operates similarly to the flash steam plant but uses the groundwater to heat a second fluid, which is then flashed. The second fluid has a

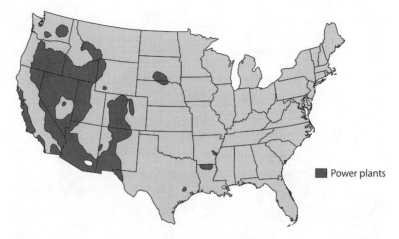

■ Power plants

FIGURE 4.7 Geothermal Resource Potential. Heat pumps for building heating and cooling are practical anywhere in the contiguous United States where local geological conditions allow pipes for the heat exchanger to be buried. The high temperatures needed to make geothermal electrical power plants practical are found only in the western United States.

Source: Adapted from EIA, "Geothermal Explained: Where Geothermal Energy Is Found," http://tonto.eia.doe.gov/energyexplained/index.cfm?page=geothermal_where.

lower boiling point than water and flashes at a lower temperature, allowing the plant to function with more widely available lower water temperatures.

Geothermal electric power plants must meet several conditions to be practical. The main condition is that the temperature must be high enough. Geothermal temperature must be at least 180°C (360°F) for flash plants and slightly less for binary plants. Such temperatures do not occur near the surface, so another condition is that the high temperature must be within drilling distance of the surface, 1 to 2 miles. The third condition is permeability of the surrounding rock. Water injected into the hot rock layer must be able to move through the rock as it is heated and extracted through the production well.

If the hot rock is dry and not permeable, it can be made permeable using an enhanced geothermal system. In this system an injection well is drilled down to the hot, dry rock, and water is pumped in under pressure. This causes the rock to fracture, opening up channels so that the water can flow through the rock and heat up. The hot water is extracted through production wells. This technology increases the area of practical geothermal energy extraction, but it is expensive.

How much electricity production can we expect from geothermal plants? At present, geothermal power provides 4% of electricity from renewable resources and 0.35% of total electricity consumption. In terms of power, geothermal provides 3,458 MW capacity, 0.3% of total generator capacity (1.1 million MW). The Department of Energy's Office of Energy Efficiency and Renewable Energy (EERE) estimates that geothermal generating capacity will increase about 15,000 MW in the next decade. EERE also predicts that most future geothermal power plants will be of the binary cycle type because the lower temperature requirements make this type of geothermal system practical over a larger area of the United States. Looking farther into the future, EERE estimates a geothermal generating capacity of 100,000 MW in fifty years with reasonable R&D and successful development of enhanced geothermal systems. Compare these numbers with the current US electricity generating capacity, 1.1 million MW. At 15,000 MW in ten years, geothermal power would be providing 1.36% of current (2010) electricity demand. At 100,000 MW in fifty years, geothermal power would be providing about 9% of current demand. However, in fifty years demand will increase markedly because of normal growth in demand and population. The switch to electric vehicles will further increase demand. Geothermal generating capacity in fifty years will probably be no more than 3% of US demand at that time.

Geothermal energy is free and essentially unlimited, though initial installation costs are high. Geothermal heat pumps can provide heating and cooling to many homes and buildings throughout the country. Direct use of hot groundwater can provide heat in some locations.

Geothermal electric power plants can contribute modestly to the country's electricity needs. Since the locations that are most suitable for geothermal power plants are in the western part of the United States, the West would benefit most.

Hydroelectric Power

Hydroelectric power comes from damming rivers and allowing the penned-up water to flow through turbines that generate electricity (fig. 4.8). In the United States, hydroelectric power accounts for 2.4% of all energy consumed, 17% of all electrical energy consumed, nearly 34% of all electrical energy from renewable resources, and 7% of total electrical generating capacity.[23] Five states—Washington, Oregon, California, New York, and Montana—produce over 70% of US hydroelectric power.

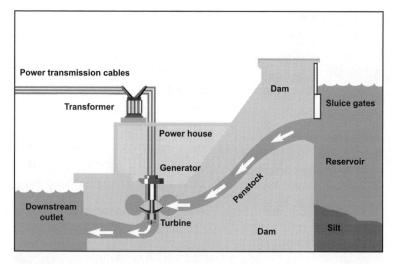

FIGURE 4.8 Hydroelectric Dam. Simplified diagram of a conventional storage hydroelectric power plant.

Source: Adapted from USGS, Water Science for Schools, "Hydroelectric Power Water Use," figure "Hydroelectric Power Generation," http://ga.water.usgs.gov/edu/wuhy.html.

In "run-of-the-river" hydroelectric systems, the normal flow of river water is through the turbine. In "conventional storage" systems, the river is dammed, and the penned up water is released through the turbine in response to varying demand for electricity. In "pumped-storage" systems, outflow from the turbine is held temporarily in a lower reservoir. Excess electricity pumps water back to the higher main reservoir so it can flow through the turbine again later, thereby "storing" electricity (though at a 15%–30% loss because of the energy needed to pump the water).

Hydropower has many of the same benefits as solar and wind power. The supply of power is essentially inexhaustible, inexpensive, pollution-free; is easily engaged and disengaged; and requires no transportation of fuel once the plant starts operation.

While hydropower is not subject to the diurnal cycle that plagues solar power or the unpredictable intermittency of both solar and wind power, it does depend on rainfall to keep the river flowing or the reservoir full. Periods of drought may seriously affect performance. Moreover, dams cause river silting, which may also permanently reduce power output. For example, silting at the Three Gorges Dam in China has severely diminished electrical output.

Hydroelectric plants require the right geological conditions. There must be a good supply of water, and the location's geology must be suitable for a dam and reservoir. The effect on the natural environment is also a consideration. Damming a river affects the environment above and below the dam, especially life in the river. The reservoir itself is large. This means that hydropower per acre is rather low, similar to solar and wind power. It also means that new dam and reservoir construction displaces large numbers of people and disrupts local communities.

Finding the right geological conditions for new hydropower plants is challenging. There is definitely a limit to how much new hydropower is available. One government study says that there are 5,677 undeveloped hydroelectric sites in the United States, with a total potential capacity of 30,000 MW,[24] a 38% increase over the existing hydroelectric capacity of 78,000 MW. Developing all 5,677 sites would increase the percentage of total US capacity provided by hydropower from the current value of 7% to 10% (fig. 4.9).

Hydropower is a clean, renewable source of electricity. Hydropower currently provides much more electricity than any other renewable resource but a small percentage of total demand for electricity. Increasing the supply of hydropower requires the correct geological conditions and consideration of the effect of a large reservoir on the environment and displaced populations. The potential increase in hydroelectric power is less than 3% of current total US generating capacity.

Clean Coal

Coal is the most polluting energy source we have. In terms of pollution per amount of energy extracted, coal produces nearly twice the carbon dioxide, seven times the carbon monoxide, about the same amount of nitrogen oxides, over twice the sulfur dioxide, forty times the particulates, and twice the mercury as oil. Considering the inefficiency of turbine generators, the actual levels of pollution from coal are three times these figures per unit of consumed electricity. Pollution and greenhouse gas emission by coal is bad now, and an increase in demand for electricity, which could be extreme if we deploy large numbers of electric cars, will drastically worsen the situation.

"Clean coal" refers to attempts to minimize CO_2 emission from coal-fired plants. There are two main approaches. Do something about pollution and CO_2 before burning the coal, or do something about CO_2 after burning the coal. The precombustion approach is complicated and requires a complete power plant redesign. In the integrated gasification combined cycle (IGCC) process, coal is first gasified and converted into carbon monoxide and hydrogen (plus the contaminants, which are removed) by reaction with oxygen and steam at high temperature.[25] This mixture of carbon monox-

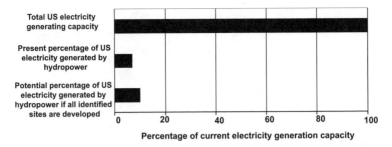

FIGURE 4.9   Potential US Hydroelectric Power. Full development of identified potential hydroelectric power sites in the United States would increase the proportion of our electric capacity from the current 7% to maximum potential of 10%.

Sources: Potential: EERE, "Wind and Water Program: Hydropower Resource Potential," www1.eere.energy.gov/windandhydro/hydro_potential.html. Current capacity: EIA, "Existing Capacity by Source," December 18, 2007, www.eia.doe.gov/cneaf/electricity/epa/epat1p2.html.

ide and hydrogen, called synthesis gas or *syngas*, is then burned to produce energy. One advantage of the IGCC process is that the syngas may be converted into CO_2 and hydrogen before being burned, and the CO_2 can easily be drawn off and processed as in carbon capture and sequestration (CCS), a postcombustion method of reducing emissions from coal. The high cost of IGCC is the biggest obstacle to its integration in the power market.

The postcombustion approach addresses CO_2 emissions and is quite simple in concept. CCS involves capturing the CO_2 produced by the power plant, compressing it to a liquid state, and then storing it. The leading current plan is to inject it into natural gas field pockets for permanent storage.

While several IGCC plants are operating in the United States and abroad, nobody has yet demonstrated the CCS process. As of mid-2008, no plant was operating CCS.[26] US efforts have stopped. The Department of Energy was funding a pilot plant but, after seven years of planning, pulled financial support when the price became too high.

Capturing and compressing CO_2 requires energy. Fuel needs of a coal-fired plant would increase between 25% and 40% with CCS. These and other system costs will increase the cost of energy from a new power plant with CCS by between 21% and 91%. That is, CCS would increase demand for resources and nearly double the price consumers pay for electricity.

My main concern about clean coal is how CO_2 is sequestered. One natural reservoir of CO_2, Lake Nyos in Cameroon, suddenly emitted a large cloud of CO_2 on August 21, 1986, suffocating seventeen hundred people and thousands of livestock. Not only is CO_2 a greenhouse gas and the major contributor to global warming, it also has a deadly potential for suffocation. The danger and the need to keep CO_2 bottled up never go away.

The lack of public reaction to the need to store CO_2, compared with the public outcry over storing nuclear waste material, bemuses me. Encapsulated nuclear waste poses no significant danger to nearby people. There are currently plans to store nuclear waste from around the country in a carefully selected and prepared repository in Yucca Mountain, Nevada.[27] Its dangerous half-life is several thousand years, but in order to cause trouble the encapsulated waste has to be broken apart and the radioactive material released.

In contrast, carbon dioxide retains its noxious properties forever. Release of a large enough amount could have a disastrous effect on global climate. Release of moderate amounts could suffocate all life in the vicinity as happened at Lake Nyos. Safety of the planned procedure to store CO_2 in numerous natural caves and caverns is far from assured.

Fischer-Tropsch Process

Another way in which we can decrease our demand for foreign oil is by obtaining gasoline from other sources, such as coal or natural gas. Both contain hydrocarbons and are possible sources of gasoline. The Fischer-Tropsch process allows making petroleum products such as gasoline and diesel from coal and natural gas.[28] Germany and Japan used the process (invented in Germany in 1920) in World War II to produce fuel because they lacked access to petroleum. The Sasol Company in South Africa currently uses this process for the same purpose, focusing on diesel. Shell is marketing synthetic fuels in Europe. In the United States, there are several demonstration projects, including one by the US Air Force in the hope of obtaining half of its aviation fuel from synthetic sources by 2016.

On the positive side, it is clearly feasible to obtain gasoline and diesel fuels from coal and natural gas, which would be a great benefit in reducing our dependence on foreign oil. Synthetic diesel has very low sulfur content, so it is superior to normal diesel with regard to pollution. On the negative side, the process requires a lot of energy and produces large quantities of CO_2, as much 7 metric tons of CO_2 per metric ton of synthetic petroleum.

It seems plausible that similar processing could produce all of the products commonly extracted from petroleum. If this turns out to be practical, then we would truly be free of the need for oil. Unfortunately, the driving force behind the Fischer-Tropsch process has always been the production of liquid fuels, and I have not found anything in the literature addressing the production of feedstock for the petrochemical industry.

At present, the relative amounts of petroleum products available in the United States depend on the refining regimen and on the mix of crude oil types input to refineries. International trade is required to maintain the desired distribution. Our continuing demand for nongasoline products such as fuel oil and petrochemical feedstock severely limits any benefit we might gain from reducing gasoline consumption. If the Fischer-Tropsch process allows us to include domestic coal and natural gas as sources of products we now derive from petroleum, the United States would be much more flexible. We should be looking at Fischer-Tropsch very carefully.

Summary

We want to reduce consumption of gasoline and diesel in order to conserve domestic reserves of oil and reduce our dependence on foreign imports. Ongoing efforts to make vehicles fuel-efficient are helping, but cannot accomplish all our goals. Replacing gasoline and diesel with nonpetroleum fuel is necessary. Research into biofuel, either bioethanol or biodiesel, is ongoing. Both suffer from low yields, which limit the potential supply of biofuel and adversely affect world food supply. In addition, the corrosive nature of ethanol poses problems for nonautomotive engines. Successful development of algae-based biofuel could resolve these issues, but we have a long way to go. As discussed in the chapter 3, switching from internal combustion engines to electric motor propulsion is an obvious path into the future. However, we need to double or triple the supply of electricity in order to meet increasing demand created by the growing fleet of electric vehicles and normal growth. This is a tall order, made more complicated because we need to do so while simultaneously reducing greenhouse gases and pollution, without damaging the environment, and without increasing consumption of limited raw natural resources.

Increased use of coal is an option. On the one hand, coal is a highly developed technology, coal is abundant, and coal is domestic. On the other hand, coal is very polluting, and mining coal is dangerous and damaging to the environment. We may be able to solve these problems, but we have a long way to go to make clean coal successful, and I am concerned about the safety of sequestering carbon dioxide for eternity. Moreover, if coal were used to meet all of the expected growth in demand for electricity, demand for coal would increase between fourfold and sixfold, presenting us with a massive undertaking.

Renewable sources of electricity—wind, solar, hydro, and geothermal power—offer a solution, but only a partial solution. Geothermal energy is free and essentially unlimited, though initial installation costs are high. Geothermal heat pumps can provide heating and cooling to many homes and buildings throughout the country. Direct use of hot groundwater can provide heat in some locations. At the national level, geothermal electric power plants can contribute modestly to the country's electricity needs, primarily in the West, where geothermal energy is most abundant. Nonetheless, good locations are limited. Today, geothermal energy provides 2.6% of our total consumption of electricity. Increasing geothermal electricity threefold would do little more than match a threefold increase in demand. Consequently, geothermal power plants probably can provide no more than 3% of future total national demand. This estimate may be overly conservative, since geothermal reserves are similar to fossil fuel reserves in that more become available as research and exploration allow us to tap into undiscovered or undeveloped reserves. However, it is doubtful that geothermal energy will provide more than a modest portion of our demand.

Hydropower is a clean, proven renewable source of electricity and currently provides a modest fraction of US demand for electricity. Hydropower currently provides 17% of our electricity, primarily in the Northwest. However, increasing the supply of hydropower requires building dams, which requires the correct geological conditions. Moreover, the large reservoir needed for a major hydroelectric dam has a large negative effect on the environment and displaced populations. The requirements for new hydroelectric power plants are well known and the estimates of potential expansion of hydropower are solid. Potential increase in hydroelectric power is approximately 50%, which would raise the percentage of national consumption of electricity provided by hydroelectric power from the current 17% to 25%. However, with a tripling of demand in the future, hydropower's contribution would be less than 10% of total national demand.

Wind power, at the scale needed to provide power to the national grid, requires good wind, large wind turbines, and lots of land devoted to wind farms. Adequate wind is found in a few areas of the United States, but the best locations for wind farming are on the Great Lakes and offshore. Noise from large turbines is a health problem that will most likely keep the land supporting wind farms from being used for anything else and will probably keep large wind farms at least several miles from areas of dense population. Offshore installations would provide reliable strong wind and minimize noise and visual pollution. If we assume wind farms can provide 100 MWh/acre/year, over 14,000 square miles of offshore wind farms would be needed to supply 22% of our current electricity demand. Devoting an 8-mile-wide band along 1,700 miles of both coasts to wind power would not provide more than 4.4% of the current demand. With the expected increase in demand, wind power will not be able to provide more than a small percentage of our demand.

Wind power is unpredictably sporadic, with occasional periods of wind dropouts during which electrical output is zero. These dropouts interrupt service to the customers but also imperil stability of the power grid. It is unlikely that grid stability can be maintained with hundreds of sporadic generators, so the output from each wind farm will have to be constant. This would require either a conventional backup generator, which would be very inefficient because the backup would be idling most of the time, or a system of backup battery storage, which would increase cost immensely. Nevertheless, battery backup, such as the NaS battery now used at some wind farms, is a feasible solution, as wind dropouts are usually rare in areas of good wind and are of short duration.

Solar electric power for the individual is simply too expensive to be practical if the national grid is available. It might be attractive to people more interested in reducing pollution and greenhouse gases than in reducing cost, but the sensible economic choice is to ignore solar power until prices come down much more than we can reasonably expect soon. At commercial power-utility scale, solar power is preferable to wind power in that the footprint in relation to power produced is smaller, 300 MWh/acre/year. A 5,000-square-mile solar farm in the Southwest (where the sun is best) could provide 22% of our current demand. Moreover, noise and visual pollution are insignificant. Countering this, periods of no sun are much more frequent and longer than periods of no wind. Unpredictable solar dropouts occur during bad weather, which might last several days, and predictable dropouts occur daily because of the diurnal cycle. Because both kinds of solar dropouts last longer than wind dropouts, battery backup is much less feasible for solar than for wind power. Because of the diurnal sun cycle, a solar farm produces electricity for only eight hours out of a twenty-four-hour day. However, US demand for electricity at night is about half the demand during the day. That is, we use as much electrical energy during the sixteen hours when the sun is not shining as we do during the eight hours of sunlight. Providing continuous output is a challenge. One potential solution is CSP systems, in which sunlight heats a fluid that drives a turbine rather than generating electricity directly in PV cells. The heat energy collected in the hot fluid by CSP systems can be stored for nighttime use.

Where does this leave us? We need to expand our supply of electricity. Renewable sources are inexhaustible and clean and do not require imports. Unfortunately, they produce limited amounts of electricity. Hydroelectric power is limited to rivers that can be dammed, geothermal power is limited by the availability of heat, power production from wind is limited to areas of good wind and sufficient land area, and solar power is limited by available sunlight and required land area. Even if we solve the practical challenges of what to do when the sun goes down or the wind dies, the renewables can provide only a limited amount of electricity. They cannot supply the majority of our electricity now and will definitely not be able to do so in the future if electric vehicles become commonplace. That is not to say that we should not develop renewable resources. We should; the benefits are clear and significant. However, we must recognize that we have to develop other sources of electricity such as nuclear power to provide what renewables cannot.

CHAPTER 5

Conclusions

W hat started as an investigation into green alternative vehicles quickly grew into an examination of energy and pollution, both traditional pollution and greenhouse gases, in general. Developing a successful vehicle that allows us to replace gasoline with a clean abundant fuel will affect all segments of the national energy system, and one cannot discuss automobiles without discussing the larger issues. In the preceding chapters, I have examined conventional and alternative energy sources as well as conventional and green vehicles. What follows is my take on the things that may work and the things we need to do to make them possible as we develop a national energy strategy. These are the things we should support with research and development (R&D) funding. You may interpret the information I have presented differently and reach different conclusions about where we should focus our R&D funding. Feel free to do so. Just let the facts, rather than the hype, guide you.

Energy

The United States and the world at large have gotten into a situation from which extrication will be difficult. After a century of ready availability of oil and gasoline, we have gone down a path that has boxed us in. Gasoline is a magnificent fuel for mobile machines and vehicles. It packs a lot of energy for its storage size and weight, is easily transported and stored, and has been, until recently, abundant and cheap. Easy and cheap transportation of people and goods has shaped worldwide behavior. At current rates of production and with increasing consumption resulting from growing populations and growing industrialization by more and more countries, the window of opportunity for action before fossil fuels run out is decades rather than centuries.

The United States imports over half of its oil, and our proved reserves are a miniscule fraction of world oil reserves. Our proved reserves will last only seven years at current rates of production, consumption, and importation. Halting imports of foreign oil by increasing domestic production does not seem to be possible, as that would require tripling production, which would be difficult. Even if it could be accomplished, our reserves would then last only three years. Our only hope for reducing greenhouse gas emissions from oil and reducing foreign imports is to reduce consumption, which is the goal of engineering green vehicles. Reducing oil consumption is more complicated than developing alternative vehicles, however. Green vehicles will reduce oil consumption if they use some other fuel, but they will reduce greenhouse gases only if the replacement fuels produce fewer emissions. Finally, one must remember that oil provides critical non-automotive-fuel products such as plastics and pharmaceuticals. Until we find replacements or other sources for these, we will need oil.

The United States imports 66% of its oil, 16% of its natural gas, and 80% of its uranium. Moreover, the United States has very little of the world's proved reserves of these resources: 2% of the oil, 4% of the natural gas, and less than 6% of the uranium. US proved reserves will not last long at current rates of production, consumption, and imports: seven years for oil, ten years for natural gas, and two hundred years for uranium. The situation with oil is even worse because US production is declining. If this trend continues, US reserves will last longer, but imports will have to increase faster to offset declining production. Reserves of oil and natural gas will probably last longer than I have stated because of undiscovered technically recoverable reserves (UTRR). The most optimistic estimates might extend sixfold the time we have before the United States runs out.

The situation with natural gas is better. The United States imports natural gas but not as large a percentage of consumption as oil. Indeed, increasing production 20%, which would be tough but feasible, would allow us to stop importing natural gas. However, in that case our domestic reserves would last only ten years. Since natural gas is cleaner than oil, it is a possible fuel for green vehicles, but using it as such would lead to increased and protracted demand, which cannot be satisfied by domestic reserves. It is doubtful that we can stop importing natural gas.

Nuclear power presents an interesting situation. Domestic reserves of uranium will last over two hundred years at current rates of production, consumption, and importation. However, we import almost 80% of our uranium. We could stop importing uranium if we were able to increase production more than tenfold, which is not possible, and even if it were, it would exhaust our domestic reserves in less than twenty years. Nuclear power provides electricity, and demand for electricity will double or triple soon in response to the demands of a growing population, the proliferation of Internet-related electronics, and the development of electric vehicles. It is doubtful that we can stop importing uranium.

Coal is also interesting. We have a lot of coal, more than any other country. Indeed, the United States exports coal, and our domestic reserves should last well over two hundred years, even if we keep exporting coal. Coal currently provides about half of our electricity, and as demand for electricity grows, domestic coal is a straightforward means of increasing generating capacity. Unfortunately, coal-fired power plants produce a large quantity of pollution and greenhouse gas emissions. Unless clean coal technology is developed successfully, replacing gasoline vehicles with electric cars powered by coal-fired power plants would produce a huge increase in noxious emissions. Even if coal is not an acceptable source of electricity, it can be used as a source of gasoline and similar fuels through the Fischer-Tropsch process. Production of gasoline from coal could help reduce imports of oil, but it would be expensive, and greenhouse gas emissions would inevitably increase.

If our abundant domestic coal is to come to the rescue, we must first get control of coal emissions, improve mining techniques to limit environmental damage, develop techniques for extracting fuel from coal, and figure out how to obtain the nongasoline petroleum products we rely on. Now, the dominant source of noxious emissions is coal-fired electric power plant turbines. Emissions will double or triple with the expected increase in demand for electricity. We should pursue efforts to reduce greenhouse gas emissions from coal. The first step, removing CO_2 before burning the fuel in an integrated gasification combined cycle plant turbine, has been demonstrated. The second step, carbon collection and sequestration, which involves capturing carbon dioxide and storing it underground permanently, has not been successfully demonstrated. The challenges of developing clean coal are daunting. Success is uncertain and, if successful, the cost will be high; clean coal will double the price of electricity. Safety is questionable; the suffocation and global warming potential of CO_2 lasts forever. I have trouble with the idea that we can store half of all the CO_2 we produce from now to eternity in hundreds of underground repositories forever. By comparison, storing comparatively small amounts of highly radioactive waste from nuclear reactors, with a dangerous half-life measured in thousands of years, in a repository we spent decades selecting and preparing, is easy.

The worldwide situation is slightly better. Proved worldwide reserves of oil should last forty years, natural gas sixty years, and uranium eighty years. The relatively small US domestic reserves are not an issue beyond their role in assuring that we will remain dependent on foreign sources. But the aforementioned projections assume that rates of consumption will remain constant. Any increase in consumption will decrease how long the reserves last. For example, an annual increase in consumption of 3%, the growth rate of US demand for oil over the past hundred years, would decrease how long worldwide proved reserves last to a couple decades.

These estimates of how long reserves will last are admittedly conservative, as they are based on proved reserves and ignore UTRR. The problem is that we don't really know how large the UTRR are, and it would be foolhardy to depend on rosy guesses provided by industry. We could start having problems in a generation, or we could escape shortages for several generations.

Green Vehicles

What about a narrower goal of eliminating, or markedly reducing, gasoline consumption? I believe that we can meet this goal with green vehicles, but trying to do so has important implications for national energy strategy. Eliminating demand for gasoline is necessary if we want to eliminate demand for oil. It will not solve the problem, as we would still have to replace the other petroleum products. Nonetheless, reducing demand for gasoline would reduce greenhouse gas emissions and reduce pressure on oil imports. Conventional wisdom places great emphasis on developing green highway vehicles, but highway vehicles are not the only consumers of gasoline. Gasoline is a light, easily stored and easily transported fuel that packs a lot of energy in a small space. There are applications that depend on these attributes much more than road vehicles. Small boats, small airplanes, motorcycles, lawnmowers, garden tools, home standby generators, and off-highway military and exploratory expeditions all require small engines and efficient fuel storage. Vehicles that operate far off the beaten track, away from fueling stations, need an easily stored and transported fuel. We will still need gasoline for these applications long after we have converted the majority of on-road automobiles to other fuels.

The current collection of fuel-efficient standard gasoline cars improve fuel economy and reduce gasoline consumption. There is some room for additional improvement, but I believe we are approaching the point of diminishing returns. Gasoline economy of 35 to 40 mpg is probably as good as we can expect. However, we must keep in mind the logistics of improving fuel economy. The recently proclaimed 35 mpg CAFE standard will not become an onroad reality until new fuel-efficient models replace all the less efficient vehicles now on the road. Complete replacement will take about forty-five years. It will be 2055 before the average fuel economy of vehicles on the road comes even close to 35 mpg, reducing demand for gasoline 43%. That is, in forty-five years we will not even cut consumption in half. Simply improving the internal combustion engine will not solve the problem.

Switching to an E85 ethanol-gasoline mixture for use in flex-fuel vehicles could reduce gasoline consumption to 15% of what it is now, but only if every vehicle were converted to E85 and supplies of ethanol were great enough. Ethanol comes from growing plants. Currently corn provides the best yield per acre, but devoting arable land to corn for ethanol takes the land out of food production. Significant corn-ethanol harvests would have an unacceptable effect on world food supplies. Sources other than corn, such as cellulosic crops, would not affect food supplies, but yields are much lower, and the supply of cellulosic ethanol would be much smaller.

Engines and fuel systems have to be modified to burn E85. If only E85 were available, all portable power tools, snowblowers, and similar small engines would have to be modified, at great inconvenience and expense. To avoid modifying every small engine, gasoline has to be available in addition to E85. In that case, since E85 is more expensive than gasoline and fuel economy is poorer, many drivers would choose gasoline instead of E85. Raising the automobile engine's compression ratio could prevent this by simultaneously improving fuel efficiency by taking advantage of the higher octane rating of E85 and dissuading drivers from filling up with gasoline. Unfortunately, it also would mean that the supply of E85 would have to be very robust to compensate for the fact that flex-fuel vehicles would no longer have the option of switching to E10.

Diesels have become clean, quiet, and fuel-efficient. New diesels get superior fuel economy. However, because of the nature of the crude oil refining process, a barrel of oil produces less diesel than gasoline. Even though diesels get more miles per gallon of diesel fuel, they get fewer miles per barrel of oil. They could actually increase the demand for oil if the number of diesels on the road increases sharply.

Biodiesel fuel could offset the increase in demand for oil, making diesels more attractive. Unfortunately, yields of the more common sources of biodiesel are so low that biodiesel cannot be the mainstay of the national transportation system. Estimates show that one should not expect much more than a few percent of total vehicle fuel from soybean biodiesel. This is certainly a step in the right direction but is hardly the entire solution. We can extract biodiesel from other crops, but the US climate favors soybeans, which have low yield. We can also import crops and biodiesel from foreign countries, but this runs contrary to becoming independent of foreign sources of energy. Biodiesel will not have a major effect on fuel consumption without a massive impact on food supplies and without our becoming dependent on foreign sources of biodiesel. This could change if algal biodiesel becomes as practical and productive as some people believe is possible. This is an intriguing possibility and something that should be closely monitored and nurtured.

Natural gas vehicles (NGVs) eliminate the need for gasoline; they are clean, and they are already successful. However, domestic supplies of natural gas are limited, and extensive conversion to NGVs would deplete our reserves faster. The United States would soon be as dependent on foreign natural gas as it now is dependent on foreign oil. In addition, the natural gas pipeline infrastructure is inadequate for widespread use of natural gas vehicles, and we would have to expand it at great effort and expense. Moreover, on-vehicle storage of natural gas requires larger and heavier fuel tanks than gasoline, resulting in conflict between the need for smaller, lighter cars and the public's demand for good driving range. For these reasons, I believer that the appropriate role for natural gas is in fleets of larger vehicles like buses, delivery vans, service vans, and the like that are fueled at depots. Fueling at central depots minimizes the need for expanded infrastructure, and deployment on large vehicles minimizes the storage tank size/weight penalty. Using natural gas in large vehicles is the most efficient application of this resource in vehicles.

Gasoline hybrid electric vehicles (HEVs) are getting very good fuel economy, around 50 mpg. Most current HEVs burn gasoline, though there is no reason they could not be powered by diesel or flex-fuel engines. If all cars on the road achieved 50 mpg, gasoline consumption would drop more than 50%, just about matching current imports of foreign oil. However, having two almost independent drive systems makes hybrids overly complex and expensive for their benefit. They are transitional between internal combustion engine vehicles and electric-drive vehicles. Switching to pure electric cars would be beneficial. Their complexity and cost would be less than HEVs', pollution would be less, consumption of oil would be less, and the cost of fuel per mile would be less.

Following the HEV as a second transitional step is the series electric vehicle (SEV), a pure electric-propulsion vehicle with electricity coming from an onboard genset—a motor/generator combination—burning fuel. The SEV would provide all the benefits of electric drive and would get maximum benefit from all of the alternative fuels such as natural gas, diesel, and biofuel. This vehicle could be the entire solution to eliminating gasoline should a nonpetroleum fuel such as algal biodiesel or coal-derived synthetic fuel become available. Especially important, we could deploy large numbers of SEVs immediately, as no new infrastructure or technology is required.

The next step would be the elimination of fossil fuel entirely by removing the SEV's fossil-fuel genset. Electric-drive vehicles are more efficient than internal combustion engine vehicles, produce no pollution themselves, and eliminate demand for oil and natural gas. That is not to say that the two leading contenders, the pure plug-in electric vehicle (P-PEV) and the hydrogen fuel-cell vehicle (HFCV), do not present challenges.

The P-PEV is similar to the gasoline-electric vehicle, but it does not generate electricity onboard. It has a large battery charged from the power grid. It is impractical now because there is no convenient charging infrastructure. Moreover, the size/weight penalty of the large battery limits driving range too much. The Chevrolet Volt is advertised at 40 miles per charge, the Nissan Leaf 100, and the Tesla Roadster 200. All have shorter range than the 320 to 400 miles per fueling stop we have come to expect. Achieving sufficient improvement in range through battery R&D is unlikely.

An alternative to storing electric energy in large batteries charged from the grid is storing hydrogen onboard and generating electricity in a fuel cell. The HFCV is impractical now, but it looks like we are on track for a practical model in a decade or so. We will then need to provide hydrogen at filling stations. Building a network of hydrogen pipelines is grossly expensive and unnecessary. One alternative would be to tap into the expanded network of natural gas pipelines that would be developed to support NGVs and convert natural gas to hydrogen at each filling station. However, it would be much more efficient to burn the natural gas in an NGV than to convert it to hydrogen for use in fuel cells. It would make more sense to manufacture hydrogen at each filling station by electrolysis. The same expanded electric grid that would support P-PEVs could support HFCVs.

Neither the P-PEV nor the HFCV is ideal. The energy storage size/weight penalty is more severe for the P-PEV, limiting range. While range is better, overall fuel efficiency of the HFCV is poor because of the losses incurred by the circular process of consuming electricity to generate hydrogen to generate electricity. We will most likely see large numbers of each type of electric vehicle. The P-PEV could provide local transportation when range is not crucial, and the HFCV could provide long-distance travel. We would just have to accept poorer fuel economy in exchange for better range, and even then the driving range of HFCVs would be less than what we expect from gasoline cars.

Deployment of electric cars will require a network of fast, very-high-power charging stations. The backbone of this network is already in place; electricity is available at all gasoline filling stations. However, electric cars will almost double demand for electricity. If increased demand from growth in population and usage of electronic devices is considered, overall demand for electricity will soon triple. We would have to increase the capacity for electric current at filling stations and feeder grids, and we would have to double generating capacity. Doubling generator capacity presents its own set of challenges. Moreover, batteries cannot accept the large charging currents required for really fast charging without damage, so charging a P-PEVs will probably always take much longer than fueling a gasoline vehicle.

Once we have converted the transportation fleet from gasoline to electricity, we will have to focus on increasing range. Battery improvement is one obvious path. Another path is making vehicles smaller, lighter, and more streamlined. Once the internal combustion engine gives way to electric propulsion, aerodynamic drag, tire rolling resistance, and inertia become the main sources of energy loss. Making the car smaller, lighter, and more streamlined provides much more benefit for an electric car than doing the same for an internal combustion engine vehicle. What we cannot achieve with battery improvement we may achieve with body design. However, making the city streets and highways safe for small cars will then become an important challenge.

Green Energy Sources

I expect demand for electricity will increase markedly because of population growth, increasing demand for electronic gadgets, and electric cars. Transmission lines and generating capacity will both have to double or triple within two or three decades. Expanding the transmission line network will face strong public resistance. Coal-fired power plants could provide the electricity, but with unacceptable increases in greenhouse gases, pollution, and environmental damage. Clean coal technology might overcome the greenhouse gas and pollution problems, but I believe clean coal has a low probability of success. I am not ready to abandon it now, but I think we should put more effort into other projects.

Expanding nuclear power is a viable option. Nuclear power is the cleanest mined-energy resource; it does not produce greenhouse gases or other pollution. World reserves should last a hundred years at current rates of consumption. Nuclear power does present some dangers: environmental issues associated with radioactive waste material, accidents, natural disasters, and terrorist activity. While these dangers are real and we have to address them, countries around the world have been using nuclear power for decades with no natural disasters or terrorist actions and remarkably few serious accidents. I believe the fears are overstated. Even so, while uranium will last a lot longer than oil or natural gas, it will not last forever. One of the potential benefits of nuclear power is the possibility of perfecting fast breeder reactors, which produce almost as much nuclear fuel as they consume, making fuel for nuclear power plants virtually inexhaustible. Some technical issues have to be resolved, but we have a hundred years to work them out before we run out of uranium. It seems to me that nuclear power will be the dominant source of electricity in the future.

Renewable sources of electricity—hydroelectric, geothermal, wind, and solar power—can help meet overall demand, but the role they can play is limited. The renewables have several features in common: though their technologies are initially high in cost, they become self-sufficient once established; they will not run out; they do not produce greenhouse gases or pollution; they can only be located where conditions are correct; and they have a large physical footprint.

Geothermal power plants require geologic reservoirs of high temperature within a couple miles of the surface. Estimates indicate that geothermal power plants might eventually provide 2% to 3% of current national demand for electricity. Hydroelectric power requires rivers with sufficient flow of water and suitable surrounding geology to make a dam and large reservoir practical. Estimates of potential hydroelectric sites indicate that fully developed hydroelectric potential in the United States could provide about 10% of total current demand.

Wind power is a definite option. Unfortunately, very few areas on land are capable of providing useful amounts of wind energy. There is public resistance to wind farms near population centers because of the hazardous noise levels produced by wind turbines. The logical choice is to put wind farms offshore, where the wind is much stronger and the noise will not bother anyone. Even so, the huge footprint of a wind farm is a major limitation. One would need offshore wind farms covering a band 9 miles wide from Maine to South Carolina and from the Canadian border to San Francisco to provide 22% of the current US demand. I doubt that much more than a quarter of this offshore swath is available for wind farms. There simply is not enough sufficiently strong wind in the United States to provide much more than 5% of our energy demand.

Unlike conventional, hydroelectric, and geothermal power plants, wind power is intermittent. Wind dies unpredictably, and power generation drops to zero. This is not good for the consumer, and it is very bad for the power grid. Power dropouts are serious problems for grid stability and can bring large segments of the grid down, causing widespread power outages. This has happened several times already and it will happen again. The solution to intermittent wind power is to ensure that output from each wind farm is steady and continuous. That is, each wind farm has to have a backup energy storage system similar to the uninterruptible power supply many computer users depend on. Uninterruptible power systems using sodium-sulfur batteries are in place at several wind farms and will probably become standard. Unfortunately, this is expensive.

One might expect hydropower and geothermal and wind power to provide as much as 20% of our current demand for electricity. This percentage would require almost maximum utilization of these resources. That is, should demand for electricity double or triple as expected, the contribution of these three renewables would decrease to less than 10% of total demand.

Solar power could supply much more electric power than these three sources combined. I estimate that to meet all of our current demand for electricity would require 36,000 square miles of solar farms and transmission lines. That is an area the size of Ohio, 1% of the country. To meet three times the current demand would require 100,000 square miles of solar farms and transmission lines, an area the size of Nevada. While technically feasible, this seems rather impractical.

Solar power has several features in common with wind power. Solar farms have to be located where the insolation is good, and they take up a large area. The footprint issue is not as severe as with wind farms because solar farms get about three times more power from an acre than offshore wind farms do. The location issue is not as critical because the best locations for solar farms are in the Southwest, where much of the land is barren and sparsely inhabited. The downside to putting huge solar farms in the Southwest is the long transmission lines that would be required, with the resulting demand for real estate for the lines and power loss in the long lines.

Like wind power, solar power is intermittent, because of clouds and bad weather, and each solar farm needs an uninterruptible power system. Solar power is unique in that it provides electricity only during daylight, around six to eight hours a day, and nothing at night. The United States uses almost as much power at night as it does during the day. This constant power demand is problematic for solar power because it means that nighttime power-generating capacity has to be as large as daytime capacity. Taking care of nighttime demand with conventional power plants and daytime demand with solar power is a massive power management problem. Solar plants would go offline in the evening and come online in the morning, while conventional plants would come online in the evening and go offline in the morning. This would be an impractical approach. An alternative is to ensure steady output from solar farms by increasing the size of the power backup to handle dropouts lasting eighteen hours. This is a simple solution, but extremely expensive. A possible solution to this problem may be found in concentrating solar power (CSP) technology, which uses sunlight to heat a fluid that drives electricity-generating turbines. The advantage of CSP over photovoltaic cells is that the heated fluid may store energy overnight.

The intermittent renewables, wind and solar power, present a quandary. We want to place solar and wind farms in areas of good conditions. That is, wind farms would be concentrated along the coast, and solar farms would be concentrated in the Southwest. But with concentration comes vulnerability. A single tropical storm or hurricane going up the east coast could shut down most of the coastal wind farms for days. A single blizzard in the Southwest could cause most of the solar farms to stop producing for days. Heavy reliance on power sources dependent on the vagaries of nature would be a serious mistake. We need to maintain conventional power-generating capacity, idle most of the time but ready to take over on short notice for wind and solar plants forced to shut down by severe weather.

Renewables may be able to provide substantial capacity, but they cannot provide the major portion of our current electricity needs. The role renewables play will only decrease in the future as growing demand outstrips geographically limited supply. It seems to me that there is no alternative to increased nuclear power.

Strategy

What is the overall strategy I am suggesting? First, we should do the following to prolong our oil supply:

- Increase domestic production and exploration, put fuel-efficient conventional and hybrid cars on the road, and convert fleets of large vehicles to natural gas.

- Explore possible nonpetroleum liquid fuel by investing in R&D on algae as a potential source of ethanol and biodiesel.

- Develop Fischer-Tropsch processes for producing gasoline, diesel, and other fuel oils from coal, while limiting emissions.

- Start phasing in electric-propulsion SEVs, thereby increasing fuel efficiency beyond that of HEVs and NGVs.

- Take advantage of the power flow in electric propulsion systems by developing lighter, smaller, more efficient road vehicles while addressing making the highways safe for smaller vehicles.

- Prepare for increased demand for electricity from P-PEVs, population growth, and increasing demand from electronic devices by developing renewable sources of electricity, expanding nuclear power capacity, expanding the transmission line network, installing a fast charging station infrastructure, continuing battery R&D, and addressing power grid stability.

- Work on developing nonpetroleum sources for the non-gasoline products we currently get from petroleum.

- Work on developing algae as a source of biodiesel fuel and other hydrocarbon products.

- Adjust to a two-fuel system in which gasoline remains available for small power devices and off-road operations and biofuel and electricity are available for highway vehicles.

- Continue to research clean coal technology, but take a hard look at its feasibility and the safety of sequestering large quantities of CO_2.

Second, we should do the following to expand our non-fossil-fuel sources of electricity to meet growing general demand and prepare for the switch to electric vehicles:

- Develop hydroelectric and geothermal power plants and solar and wind farms where appropriate and economical, expand the transmission line network, and develop widespread fast charging stations for electric vehicles.

- Ensure power grid stability by requiring an uninterruptible power system at each solar farm and wind farm.

- Improve power grid stability to accommodate intermittent power from solar and wind farms, thereby lessening the required capacity of uninterruptible power systems and facilitating the connection of community-level or individual solar and wind systems to the grid.

- Increase conventional electric power generator capacity by constructing nuclear power plants.

- Deal with concerns about nuclear plant safety, security against terrorist activity, and storage of waste materials and rescind the prohibition against reprocessing spent fuel rods.

- Continue investment in fast breeder reactor technology, which could provide almost inexhaustible nuclear energy.

This plan should markedly reduce our dependence on oil and markedly reduce noxious vehicle emissions. Unfortunately, it will not eliminate demand for gasoline, as we will probably still need it for the small engines in power tools and so on. Perhaps biofuel could replace gasoline in small engines, but even that will not eliminate demand for oil, as we will still need fuel oils, lubricating oils, and all the other products we get from oil. We may eventually learn how to do without or extract these products from coal, but we will need oil for a long time.

I believe the actions listed above provide a logical general strategy that accounts for all the nuances of green vehicles and eliminating dependence on oil. I make no claim that I have answered all the questions. Oil will run out. Natural gas will run out. Uranium will run out. With the uncertainty in estimates of reserves of natural resources, we do not know exactly when. Nevertheless, it looks as if our grandchildren or great-grandchildren will be living in a very different world. I hope you now have a better grasp of the issues and the facts behind the sound-bite headlines and are better prepared to address the very important energy issues that the future has in store for us all.

Conversions

LENGTH

1 meter	39.37 inches
	3.281 feet
1 km	0.6214 mile
	3,281 feet
1 mile	5,280 feet
	1.609 km

AREA

1 acre	43,560 sq. feet
	4,047 sq. meters
1 sq. mile	640 acres

VELOCITY

1 m/s	2.237 mph

VOLUME

1 L	0.26417 US gallon

ENERGY

1 quad	10^{15} Btu
	293 billion kWh
1 MJ	0.278 kWh

WEIGHT

1 kg	2.2046 pounds
1 long ton	2,240 pounds
1 short ton	2,000 pounds
1 metric ton	1,000 kg
	2,204.6 pounds

ENERGY DENSITY

1 MJ/kg	0.1260 kWh/lb
1 MJ/L	1.0516 kWh/gal

Notes

....

One: Conventional Energy Sources

1. EIA, www.eia.doe.gov/.

2. This book uses the "short scale" of numbers common in the United States rather than the "long scale," which is used in some other countries. In the short scale, 1 million is 1,000,000 or 1×10^6; 1 billion is 1,000,000,000 or 1×10^9; 1 trillion is 1,000 billion or 1×10^{12}; 1 quadrillion is 1,000 trillion or 1×10^{15}. See http://en.wikipedia.org/wiki/Long_and_short_scales for an interesting discussion of long and short scales. 1 quad = 1 quadrillion Btu = 1.0×10^{15} Btu = 293 billion kWh. (1 kWh = 3,412 Btu).

3. A metric ton = 1,000 kg, or 2,204.6 pounds, the mass of a cubic meter of pure water. A short ton = 2,000 pounds. A long ton is 2,240 pounds. Standard usage in the literature is to qualify the term *ton* as metric, short, or long. The term *ton* by itself is generally taken to mean a short ton.

4. Nongreenhouse-gas pollutants include sulfur oxides, in particular sulfur dioxide (SO_2) from industrial processes, especially combustion of coal and petroleum; nitrogen oxides (NO_x), especially nitrogen dioxide (NO_2), from high-temperature combustion such as that by internal combustion automobile engines; and carbon monoxide (CO) from incomplete combustion of fuel, especially in vehicle exhaust.

5. EIA, "Emissions of Greenhouse Gases Report," December 3, 2008, www.eia.doe.gov/oiaf/1605/archive/gg08rpt/index.html.

6. Intergovernmental Panel on Climate Change, www.ipcc.ch/.

7. David A. Fahrenthold, "Chemicals That Eased One Woe Worsen Another," *Washington Post*, July 20, 2009.

8. Andrew Higgens, "A Climate Threat, Rising from the Soil," *Washington Post*, November 19, 2009.

9. For interesting background see Wikipedia, "Petroleum," http://en.wikipedia.org/wiki/Petroleum.

10. For interesting background see Wikipedia, "Oil Refinery," http://en.wikipedia.org/wiki/Oil_refining.

11. EIA, "Petroleum Navigator," Definitions, Sources, and Explanatory Notes, http://tonto.eia.doe.gov/dnav/pet/TblDefs/pet_crd_pres_tbldef2.asp (emphasis added).

12. MMS, "Assessment of Undiscovered Technically Recoverable Oil and Gas Resources of the Nation's Outer Continental Shelf, 2006," www.boemre.gov/revaldiv/PDFs/2006NationalAssessmentBrochure.pdf.

13. USGS, "USGS National Assessment of Oil and Gas Resources Update (December 2007) Conventional Oil and Gas Resources," http://certmapper.cr.usgs.gov/data/noga00/natl/tabular/2007/summary_07.pdf.

14. MMS, "Assessment of Undiscovered, 2006," cited above (emphasis added).

15. USGS, "3 to 4.3 Billion Barrels of Technically Recoverable Oil Assessed in North Dakota and Montana's Bakken Formation—25 Times More Than 1995 Estimate," www.usgs.gov/newsroom/article.asp?ID=1911.

16. EIA, "Independent Statistics and Analysis," Basics tab, http://tonto.eia.doe.gov/energyexplained/index.cfm?page=coal_home#tab1.

17. Ibid., Data & Statistics tab, http://tonto.eia.doe.gov/energyexplained/index.cfm?page=coal_home#tab2.

18. "*BP Statistical Review of World Energy June 2007.*" The data are also found in readable format at Wikipedia, "Table: Coal; Proved Recoverable Coal Reserves, at end-2006," http://en.wikipedia.org/wiki/Coal. Note that the data have been converted from tonnes to short tons.

19. EIA, *Annual Energy Review, 2009*, table 9.2, "Nuclear Power Plant Operations1957–2009," www.eia.doe.gov/emeu/aer/pdf/pages/sec9_5.pdf.

20. World Nuclear Association, "World Nuclear Power Reactors and Uranium Reactors," November 1, 2010, www.world-nuclear.org/info/reactors.html.

21. World Nuclear Association, "Uranium Production Figures, 1999–2009," July 2010, www.world-nuclear.org/info/uprod.html.

22. World Nuclear Association, "World Uranium Mining," updated May 2010, www.world-nuclear.org/info/inf23.html.

23. World Nuclear Association, "Safety of Nuclear Power Plants," updated September 13, 2010, www.world-nuclear.org/info/inf06.html.

24. Nuclear Regulatory Commission, "Backgrounder on the Three Mile Island Accident," www.nrc.gov/reading-rm/doc-collections/fact-sheets/3mile -isle.html (emphasis added).

25. World Nuclear Association, "Chernobyl Accident," updated August 2010, www.world-nuclear.org/info/chernobyl/inf07.html (emphasis added).

Two: Conventional Vehicles

1. Erik Eckermann, *World History of the Automobile*, trans. Peter L. Albrecht (Warrendale, PA: Society of Automotive Engineers, 2001). For additional background, see Wikipedia, "History of the Automobile," http://en .wikipedia.org/wiki/History_of_the_automobile, and Wikipedia, "Automobile," http://en.wikipedia.org/wiki/automobile.

2. Bureau of Transportation Statistics, table 1-11, "Number of U.S. Aircraft, Vehicles, Vessels, and Other Conveyances," www.bts.gov/publications/ national_transportation_statistics/html/table_01_11.html.

3. Bureau of Transportation Statistics, table 4-23, "Average Fuel Efficiency of U.S. Passenger Cars and Light Trucks," www.bts.gov/publications/ national_transportation_statistics/html/table_04_23.html.

4. Richard Stone and Jeffrey K. Ball, *Automotive Engineering Fundamentals* (Warrendale, PA: SAE International, 2004), 20–32.

5. Stone and Ball, *Automotive Engineering Fundamentals*.

6. DOE, "Energy Efficient Technologies," www.fueleconomy.gov/feg/ tech_adv.shtml.

7. EPA, "EPA's Fuel Economy and Emissions Programs," www.epa.gov/ fueleconomy/420f04053.htm#calc.

8. NHTSA, "CAFE Overview—Frequently Asked Questions," www .nhtsa.dot.gov/CARS/rules/CAFE/overview.htm; Robert Bamberger, "Automobile and Light Truck Fuel Economy: The CAFE Standards," *Almanac of Policy Issues* (September 2002), www.policyalmanac.org/environment/ archive/crs_cafe_standards.shtml; Wikipedia, "Corporate Average Fuel Economy," http://en.wikipedia.org/wiki/Corporate_Average_Fuel_Economy.

9. DOE, "How Vehicles Are Tested," www.fueleconomy.gov/feg/how_ tested.shtml.

10. Peter Whoriskey, "GM Says New Car Is Capable of 230 MPG," *Washington Post*, August 12, 2009.

11. DOE, "Gas Mileage Tips," www.fueleconomy.gov/feg/drive.shtml.

12. Michael Booth, "Hypermilers Stretch Their Gas Mileage," *Denver Post*, July 9, 2008.

Three: Green Vehicles

1. Greg Pahl, *Biodiesel: Growing a New Energy Economy*, 2nd ed. (White River Junction, VT: Chelsea Green, 2008).

2. For example, Pahl, *Biodiesel*.

3. John Sheehan et al., *A Look Back at the U.S. Department of Energy's Aquatic Species Program: Biodiesel from Algae*, NREL/TP-580-24190 (Golden, CO: NREL, 1998), www1.eere.energy.gov/biomass/pdfs/biodiesel_from_algae.pdf.

4. Algenol Biofuels, "BioFields," www.algenolbiofuels.com/biofields.htm.

5. Jim Lane, "Biofields to Commence Construction in January on Algae-to-Ethanol Pilot Using Algenol Technology," January 6, 2010, http://cleanenergysector.com/2010/01/biofields-to-commence-construction-in-january-on-algae-to-ethanol-pilot-using-algenol-technology/.

6. EthanolRetailer.com, "Find an E85 or Blender Pump Station at Growth Energy's PumpFinder," www.e85fuel.com/find-an-e85-station/.

7. EPA, "Fuel Economy Guides, 2000–2011," www.fueleconomy.gov/feg/FEG2000.htm.

8. Wisconsin Department of Commerce, "Preparing Your Dispensing System for Ethanol Blended Motor Fuel," www.commerce.state.wi.us/ER/pdf/bst/ProgramLetters_PL/ER-BST-PL-PreparingForEthanolBrochure.pdf.

9. David Korotney, "Water Phase Separation in Oxygenated Gasoline," memorandum to Susan Willis (EPA, 1995), www.epa.gov/oms/regs/fuels/rfg/waterphs.pdf.

10. "Ethanol Exemption Becomes Law in Oregon," *BoatU.S. Magazine*, May 2008, 7.

11. Pressure Solutions LLC, "Fill Up Your NGV at Home with Phill by FuelMaker," http://pressuresolutionsllc.com/Phill.html.

12. Annual fuel saving is $524 with fuel economy improvement from 21 mpg to 29 mpg at 15,000 miles per year and $2.66 per gallon.

13. EPA, "Fuel Economy Guide, 2009," www.fueleconomy.gov/feg/FEG2009.pdf.

14. Tesla Motors, www.teslamotors.com/; Chevrolet, "2011 Volt," www.chevrolet.com/volt/

15. EIA, "Frequently Asked Questions—Gasoline," www.eia.doe.gov/ask/gasoline_faqs.asp#retail_gasoline_stations.

16. EIA, "Existing Capacity by Energy Source," www.eia.doe.gov/cneaf/electricity/epa/epat1p2.html.

17. Battery University, "Charging Lithium-Ion Batteries," http://batteryuniversity.com/index.php/learn/article/charging_lithium_ion_batteries.

18. To be precise, a *generator* converts one form of energy into electricity. Something else provides the initial source of energy. That is, in a conventional power plant or home generator, a fossil-fuel-burning *motor* or turbine turns a shaft that drives the generator. Hence the two-stage system consisting of a motor and a generator (i.e., a genset).

19. Peter Whoriskey, "GM Says New Car Is Capable of 230 MPG," *Washington Post*, August 12, 2009.

20. For background see Wikipedia, "Lithium-Ion Battery," http://en.wikipedia.org/wiki/Lithium_ion_battery.

21. US Geological Survey minerals report, http://minerals.usgs.gov/minerals/pubs/commodity/lithium/mcs-2010-lithi.pdf; William Tahil, "The Trouble with Lithium: Implications of Future PHEV Production for Lithium Demand" (Meridian International Research, December 2006), www.evworld.com/library/lithium_shortage.pdf.

22. Tahil, "Trouble with Lithium," 6.

23. Bureau of Transportation Statistics, "From Home to Work the Average Commute Is 26.4 Minutes," *Omnistats* 3, no. 4 (October 2003), www.bts.gov/publications/omnistats/volume_03_issue_04/. Charging stations at the workplace would alleviate the problem for many commuters, but the almost 8% of commuters with 40-mile or more one-way commutes would still need fuel for part of their commute.

24. DOE, "Hydrogen Program," www.hydrogen.energy.gov/.

25. "Hydrogen Cars: Fad or the Future?" *Science* 324 (June 5, 2009): 1257–59. Average prices were $275 per kW in 2002, $110 per kW in 2006, and $73 per kW in 2008. The trend is on track to reach the $30 per kW goal by 2015. This means that the 134 horsepower, 100 kW fuel cell in the Honda FCX Clarity should cost around $3,000 by 2015. For general background on fuel cells, see Wikipedia, "Fuel Cell," http://en.wikipedia.org/wiki/Fuel_cell.

Four: Green Energy Sources

1. American Wind Energy Association, www.awea.org; NREL, www
.nrel.gov/; EIA, "Wind," www.eia.doe.gov/cneaf/solar.renewables/page/wind/
wind.html.

2. Vestas, www.vestas.com/.

3. NREL, "Wind Farm Area Calculator," www.nrel.gov/analysis/power
_databook/calc_wind.php.

4. Cape Wind, www.capewind.org/index.php.

5. Cape Wind, "View from the Cape and Islands," provides several art-
ists' renditions of how the facility will look. See www.capewind.org/modules
.php?op=modload&name=Sections&file=index&req=viewarticle&artid=9.

6. Steven Mufson, "Pickens Calls Off Plans for Vast Texas Wind Farm,"
Washington Post, July 11, 2009.

7. Pickens Plan, www.pickensplan.com/theplan/.

8. Dan Charles, "Renewables Test IQ of the Grid," *Science* 324 (April
10, 2009): 172–75.

9. Akhiro Bito, "Overview of the Sodium-Sulfur (NAS) Battery for the IEEE
Stationary Battery Committee" (NGK Insulators, June 15, 2005), www.ieee
.org/portal/cms_docs_pes/pes/subpages/meetings-folder/2005_sanfran/Non
-Track/Overview_of_the_Sodum_-_NAS_IEEE_StaBatt_12-16Jun05_R.pdf;
Y. Kishinevsky, "Long Island Bus Sodium Sulfur Battery Storage Project,"
www.sandia.gov/ess/docs/pr_conferences/2005/Kishinevsky.pdf. Interesting
background is found at http://en.wikipedia.org/wiki/NAS_battery.

10. NREL, "U.S. Solar Radiation Resource Maps," http://rredc.nrel.gov/
solar/old_data/nsrdb/redbook/atlas/.

11. Airman 1st Class Ryan Whitney, "Nellis Activates Nations Largest
PV Array," www.nellis.af.mil/news/story.asp?id=123079933.

12. Daylight in the US ranges from ten to fifteen hours per day, depending on the season. Sunlight is too dim near sunrise and sunset to provide much
energy. A good rule of thumb is eight hours per day of useful electric output
from solar panels.

13. Mr. Solar (www.mrsolar.com) suggested a CE60208 system costing
$71,500 for the Baltimore, MD, area.

14. Public Service Commission of Wisconsin, "Electric Transmission
Lines," http://psc.wi.gov/thelibrary/publications/electric/electric09.pdf.

15. NREL, "Concentrating Solar Power Research," www.nrel.gov/csp/.

16. Wikipedia, "Parabolic Trough," http://en.wikipedia.org/wiki/Trough_concentrator.

17. Wikipedia, "Solar Thermal Energy: Dish Designs," http://en.wikipedia.org/wiki/Solar_thermal_energy#Dish_designs.

18. John A. Duffie and William A. Beckman, *Solar Engineering of Thermal Processes*, 2nd ed. (New York: John Wiley and Sons, 1991); Mukund R. Patel, *Wind and Solar Power Systems* (Boca Raton, FL: CRC Press, 1999).

19. EIA, "Geothermal Explained: Use of Geothermal Energy," http://tonto.eia.doe.gov/energyexplained/index.cfm?page=geothermal_use.

20. EIA, "Net Generation by Energy Source," table 1.1, www.eia.doe.gov/cneaf/electricity/epm/table1_1.html. Also see table 1.1.A, "Net Generation by Other Renewables," www.eia.doe.gov/cneaf/electricity/epm/table1_1_a.html.

21. EIA, "Geothermal Explained: Geothermal Power Plants," http://tonto.eia.doe.gov/energyexplained/index.cfm?page=geothermal_power_plants.

22. EERE, "Geothermal Technologies Program," www1.eere.energy.gov/geothermal/powerplants.html.

23. EIA, "Existing Capacities by Energy Source" (November 23, 2010), www.eia.doe.gov/cneaf/electricity/epa/epat1p2.html.

24. EERE, Wind & Water Power Program, "Hydropower Resource Potential" (2009), www1.eere.energy.gov/windandhydro/hydro_potential.html.

25. Wikipedia, "Gasification," http://en.wikipedia.org/wiki/Gasification.

26. Wyatt Andrews, "Clean Coal—Pipe Dream or Next Best Thing?" *CBS Evening News*, www.cbsnews.com/stories/2008/06/20/eveningnews/main4199506.shtml.

27. As of this writing, after twenty years and a $10 billion investment, funding for continued development of the Yucca Mountain Repository has been withdrawn, leaving only some funding for a Blue Ribbon Commission on Nuclear Waste tasked to come up with an alternative. However, there is also congressional effort to reinstate development funding, so the future of the Yucca Mountain Repository is uncertain.

28. Wikipedia, "Fischer-Tropsch Process," http://en.wikipedia.org/wiki/Fischer-Tropsch; EPA, "Clean Alternative Fuels: Fischer-Tropsch," www.afdc.energy.gov/afdc/pdfs/epa_fischer.pdf.

Index

Chernobyl nuclear reactor, 38, 39–40
Chevrolet: Aveo, 51, 81; Cobalt, 94, 94; Equinox, 97, 99–100, 101
Chevrolet Volt: battery of, 83, 92; Cobalt compared to, 94, 94; fuel economy of, 56, 89; as H-PEV, 96, 106–7; internal combustion engine for, 88; range of, 85, 156
city driving, 46, 48, 49, 80
Clean Air Act of 1990, 69
clean coal approaches, 140–42, 151, 158
CO_2. See carbon dioxide
coal: clean coal approaches, 140–42, 151, 158; deaths due to, 38; in electric power generation, 9–10, 10, 30, 31–32; greenhouse gas emissions due to, 8, 9–10, 10; petroleum products from, 142–43; pollution due to, 11, 11, 140; potential of, 42–43, 150–51; production, consumption, and imports of, 2, 3, 5, 30, 42–43; reserves of, 30–31, 31; types of, 29–30; underground coal-seam fires, 9
compressed natural gas (CNG), 25, 74
concentrating solar power (CSP) thermal systems, 131–32, 161
conversions, 165
cooking oil as biofuel, 67–68
corn, as biodiesel crop, 66–67, 67
Corporate Average Fuel Economy (CAFE) standards, 51–54
corrosiveness of ethanol, 71–72
cost: of alternative vehicles, 62; of cars, and CAFE standard, 52–53; of clean coal approaches, 141–42, 151; of E85, 71; of HEVs, 80, 81, 82; of hydrogen fuel cells, 101; of NGVs, 75; of solar power, 128–29; of uranium, 40–41
countries: natural gas imports by, 28; nuclear power production of electricity by, 34; oil imports by, 18; proved coal reserves by, 31; proved

natural gas reserves by, 28; proved oil reserves by, 23; proved uranium reserves by, 37; sources of lithium by, 93
cracking, 13
crops for fuel, 66–67, 67, 68, 70, 154
CSP (concentrating solar power) thermal systems, 131–32, 161

dairy consumption, 8, 9
data, sources of, xii
demand: for diesel fuel, 22, 153–54; for electricity, 87, 109, 147, 157; for gasoline, xi–xii, 12, 22–23, 152–53; for natural gas, 27, 77; for oil, xvii–xviii, 16–17, 17, 43, 59, 153–54, 163–64
density of petroleum, 15
design speed of cars, 49
Diesel, Rudolf, 44, 63, 65
diesel fuel: consumption of, 6; greenhouse gas emissions due to, 8, 9, 10, 44; reducing demand for, 22; refining process and, 14, 14
diesel vehicles: biodiesel, 63, 65–68; demand for oil and, 153–54; functional layout of, 62, 64; overview of, 61–63, 68–69, 103–4. See also diesel fuel
dish CSP systems, 132
Dodge Avenger, 56
driving: in city, 46, 48, 49, 80; efficiently, techniques for, 57; on highway, 47, 48, 80
dry-steam geothermal power plants, 134–35

E10 ethanol fuel, 72–73
E85 ethanol fuel, 56, 69–73, 153
efficient driving, techniques for, 57–58
electricity: consumption of, 6, 87, 130–31; demand for, 87, 109, 147, 157; expanding non-fossil-fuel sources of, 163. See also electric

greenhouse gas emissions: from coal, 8, 9–10, *10*, 151; by energy content, 9–11, *10*; from gasoline and diesel, *8*, *9*, *10*, 44; global warming and, 6–7; by source, *8*
gridded solar systems, 126

heliostat tower CSP systems, 132
HEVs. *See* hybrid electric vehicles
HFCVs. *See* hydrogen fuel-cell vehicles
highway driving, 47, *48*, 80
home recharging of EVs, 85–87
home refueling of NGVs, 76
Honda: Civic, 74–75, *75*, 81; FCX Clarity, 97, 98
hybrid electric vehicles (HEVs), *64*, 78–82, 105, 155
hybrid plug-in electric vehicles (H-PEVs), *65*, 96, 106–7. *See also* Chevrolet Volt
hydrocarbons in petroleum, 12–13
hydrofluorocarbons, 7
hydrogen fuel-cell vehicles (HFCVs): gasoline vehicles compared to, *101*; layout of, *84*; overview of, 97–103, 107–8; PEVs compared to, *102*; potential of, 156–57
hydropower/hydroelectric power: consumption of, *3*, *4*; dam diagram, *138*; deaths due to, 38; overview of, 138–40, 145; potential for, *141*, 159–60
hypermiling, 58

idling losses, 46, 49, 79–80
importation: dependence on, 41–42; fuel economy increases and, 59; of natural gas, 25, 26–27, *27*, *28*, 77; of petroleum, 16–18, *17*, *18*, 19, 22–23; of uranium, 35
industry, consumption of energy by, *5*, *26*
integrated gasification combined cycle process (IGCC), 140–41
Intergovernmental Panel on Climate Change (IPCC), 7

internal combustion engines: in alternative vehicles, 60–61; in Chevrolet Volt, 88; in diesel vehicles, 61–69; efficiency of, 58; in flex-fuel vehicles, 69–73; improvements to, 103; layout of, *64*; losses typical in, 46–47, *48*, 49–50; in natural gas vehicles, 73–78
interstate highway system, 45

lighting and petroleum industry, 12
linear CSP systems, 132
liquefied natural gas (LNG), 24–25, 74
lithium for batteries, 92, 93, *93*

manufacturer compliance with CAFE standard, 53–54
meat consumption, *8*, 9
mercury, pollution due to, *11*
methane, as greenhouse gas, 7, *8*, 9
methyl tertiary butyl ether (MTBE), 69
Middle East, oil imports from, 17–18, *18*

national energy strategy: actions in, 161–64; alternative vehicles, 103–8; coal, 150–51; fossil fuels, 148–50; goals of, 1, 103; green energy sources, 157–61; green vehicles, 152–53; natural gas, 150; nuclear power, 150
national power grid, 87, 118–19, 127–28
National Renewable Energy Laboratory (NREL), 123
natural gas: consumption sectors, *5*; deaths due to, 38; demand for, 27, 77; greenhouse gas emissions due to, *8*, *9*, *10*; petroleum products from, 142–43; pollution due to, 11, *11*; potential for, 150; production, consumption, and imports of, *2*, *3*, 25–26, *27*, 41, 42; reserves of, 26–27, *28*; transportation and storage of, 24–25, 29. *See also* natural gas vehicles